The Meaning of Evolution

by Samuel Christian Schmucker

CONTENTS

A FOREWORD

Before my window lies an enchanting landscape. It embraces a stretch of open rolling country, beautiful as the eye could wish to rest upon. The sun with its slanting rays is not giving it heat enough in these winter months to make it blossom in its radiant beauty, but the mind goes easily back through the few brown months to the time when the field not far away was waving with its rich yellow grain so soon to be food for those who planted it. Beyond this field lies an orchard where, in regular and orderly rows, stand the apple trees whose bright blossoms in the spring make the landscape so beautiful and whose fruit in the fall serves so richly for our enjoyment. A little farther on, a pasture is filled with sleek-coated cows, feeding quietly and patiently until the evening when they will return to their stalls to yield their rich milk. Still farther on lies a tract of forest. The varied shades of the beeches, the tulip poplars and the chestnuts make an exquisite contrast and give to the landscape its attractive background framed in by a distant hill. Behind this hill flows a mighty river carrying on its breast the ships by which we share the over-abundance of our own blessings with our brothers on the other side of the sea, from whom in turn we receive of their overplus. Beyond this teeming river lies a level stretch of fertile land and then the mighty ocean. On one side of the scene runs a busy highway. Along this men pass and repass, some on foot, others drawn by their patient and submissive horses. Still others are carried by the new-found power of the sunshine imprisoned beneath the rocks in the oil that has been forming ever since the sun shone down upon the great forests of the far distant past.

In a pathway to one side, some children are playing. One of them has laid upon the ground a rectangle of stones divided into four and her little mind sees before her the house which is teaching her to get ready for the work that shall come to her in later life. Meanwhile her short-haired companion is prancing around astride a stick; he too, little as he suspects it, is getting ready for life.

It needs little reflection to realize that the scene has not always been what it is. The underlying ground has surely been there longest, its age vying only with that of the bounding ocean that beats upon the shore and works the sand into fantastic stretches. The forest has been there long and so has the stream; the road perhaps ranks next in age; then come the orchard trees, and most recent of all the waving grain. People come and go but form no stable part of this landscape. We know how the grain came to be there, and we understand the orderly arrangement of the orchard trees; the road too we can explain. How came the stream there, and how the forest trees? Have they always been there, or did they too have a beginning? Was there a time when there was no ocean? When was this time? How came they there?

When the lisping lips of my young child asked me, "Papa, who made me?" I told him "God," and he knew enough and was content with his knowledge. After a while he grew older and his inquisitive spirit began to puzzle with the question of how God had made him. When his growing mind was ready for the new knowledge I took him to my side and told him the great mystery of life. I told him how he owed to his father and to his mother the beginnings of his life, how God gave him to us. Now a new era opened in his childish mind. As he grows on to greater maturity he cannot help wondering how the first man was made, how the trees, and the world came to be. He is no longer satisfied with the simple statement that God made them. His eager mind wants to know, if may be, how God made them.

So, in the distant past, in the childhood of our race, the question was asked, "Who made us?" and the answer was "God." Men formed their simple conception at that time of how He did it. As the centuries rolled by and the children of men have learned from creation the story of its origin a riper and richer knowledge has given them a broader and finer conception. No less does the reverent student believe that God created the earth, but he no longer thinks of God as working, as man works. He no longer feels that it is impious to attempt to read God's plan in His work; to see how this work has arisen, to see, if may be, what there is ahead.

This is one of the tasks to which science is now giving itself. The answer is uncertain and halting. A few things seem clear; others seem to be nearly certain; of still others we can only say that for the present we must be content with the knowledge we have. But if we take the best we have and work over it thoughtfully and carefully, the better will slowly come, and in time we shall know far more than we now suspect. Meanwhile, it is the attempt of this book to give to people whose training is other than scientific some conception of this great story of creation. Without dogmatic certainty but without indecision it tries to tell what modern science thinks as to the great problems of life. It tries to describe the possible origin of animals and plants, their slow advance, the length of their steady uplift, the forces that brought it about. It tries to tell a little of the men who have helped to develop the great idea of evolution, of the great men who persuaded the scientific world of its truth, and of the later minds that are modifying and enlarging the idea of the master evolutionist. It tries to tell what science perhaps vaguely hopes as to the future. What are we to be? Can we help the great advance?

The Meaning of Evolution

CHAPTER I

EVOLUTION BEFORE DARWIN

Ever since men have been able to think they must have puzzled out for themselves some way of accounting for their own beginnings. Every savage tribe with whom we have any intimate acquaintance has some story that accounts for the origin of the tribe at least, and often for the beginning of the world. These stories are handed down from generation to generation and are scarcely questioned in the thought of most men. In early Greece there was a succession of men whom the world calls philosophers. These men thought earnestly and deeply on all kinds of questions. Their method was not our method. The plan of making a long series of observations, before coming to any conclusion, was not the habit of their minds. They reasoned out on general principles what seemed to them must have been the origin of the world. It is not strange that among these should come, now and then, some one who in some passage or other should show that there had come to his mind at least a glimmer of the thought that was later to develop into the great idea which the modern world calls evolution.

Among the earliest of these was Anaximander, who lived 600 years before Christ. He thought that the earth was at first a fluid. Gradually this fluid began to dry and grow thicker, and here and there, where it thickened most, dry land appeared. When this dry land had become firm enough to serve as his home, man came up from the water in the form of a fish. Slowly and gradually the fish, struggling about on the land, gained for himself the limbs and members he needed for his new situation and developed into a man. After him other animals came up in much the same fashion, then the plants, until the whole world was clothed with its present inhabitants.

One hundred and fifty years later Empedocles announced a new thought. He said that in the beginnings there were all sorts of strange, incomplete, and misjointed monsters which swarmed upon the earth, having sprung up out of the earth itself. Each was a chaos of the limbs which afterward were to belong to other animals which needed them more. Slowly and gradually an interchanging came about by which appropriate limbs fastened themselves to the proper animals. The last of these misjointed creatures is the one known as the centaur, half-man--half-horse. After a while, when all the members had found their proper places, the animals were complete. In one respect this opinion foreshadowed our later idea. It suggested that the more perfect animals had arisen out of the less perfect and that the change came gradually.

Then came Anaxagoras, who was the first to believe that there was intelligent design back of the creation of animals and of plants. He thought there had originally been a slime in which were the germs of all the later plants, animals, and minerals, mixed in a chaos. Slowly order arose. Out of the mixture settled first the minerals forming the earth, with the air floating above it, and above the air was the ether. Out of the air the germs of plants settled upon the earth, and vegetation covered the mineral floor. Then from the ether came the germs of animals and of men. These settled among the plants and sprang up into the animals of the world, as well as the people.

The greatest scientific thinker of early Greece was Aristotle. He had lived by the seashore and knew better than any other man of his times the exquisite seaweeds and the still more beautiful marine animals. He was the first to think of them as a linked series, the higher developing out of the lower under the pressure of what he called a perfecting principle. Out of the inanimate rocks had sprung the marine plants--the seaweeds. From these had developed first "plant animals" like the sea anemones and the sponges. These grew attached to the rocks, as plants do. With higher development came locomotion, with ever-increasing energy. At last man arose, the crown of all creation. Presiding over all this advance is the "efficient cause," God. Aristotle rejected entirely the earlier ideas that any of this work came about by chance. He was certain of the existence of plan and purpose in the development.

Just a little before the time of Christ the Latin poet, Lucretius, wrote a poem on "The Nature of Things." Here he describes how in the early years the beginnings of things in small, disjointed fashion moved about among each other at first in utter confusion, each trying itself with the other. After many trials the proper members came together. When they had been thus placed the warmth of the sun shining down upon the earth helped the earth to reproduce the same sort of creatures. So living things came up and flourished. The poem expresses many beautiful ideas, but the underlying conceptions lack the unity and grandeur that marked Aristotle's work, which later was the potent influence in shaping men's minds. It died out after a while, only to awake in the Renaissance with marvelous vitality, starting the world to think afresh great thoughts that would not die, but would grow from that time on with ever-widening scope.

Among the Jews and early Christians the stately and beautiful account in Genesis sufficed for all the needs of minds fully occupied with other questions. With the growth of philosophy among Christian minds again came the need of a satisfactory solution. St. Augustine was probably the greatest of the so-called "Fathers" of the church. His mind was eminently philosophical, and he was learned in the writings of the older Greeks. He believed the language of Genesis to mean that in the beginning God planted in chaos the seed that afterward sprang up into the heavens and the earth. He further says that the six days of creation were not days of time, but a series of causes, and that, in the order described as these six days, God planted in chaos the various beginnings of things. These in the fullness of time sprang up into the world as we know it now. The problem was not a question about which the church cared to trouble itself, and with the oncoming of the Dark Ages the whole matter dropped nearly out of the thoughts of men.

When the times began to lighten we find the schoolmen, among the greatest of whom was Thomas Aquinas. Referring especially to the authority of his master, St. Augustine, he says that it would be easy mistakenly to believe that the author of Genesis meant to convey the idea that on each of the six days certain acts of creation were performed. It is quite evident, thinks Aquinas, that in those early times God only created the germs of things and put into the earth powers which should later become active. After the Creator had thus endowed the earth he rested from the work, which proceeded to develop under the influence of these first germs.

Nearly four hundred years later, when Europe had finally awakened out of the deep and refreshing sleep in which it had fortunately forgotten much of the past, a new era dawned and modern thought began. Immediately men commenced to busy their minds with broader problems than they had been discussing since the time of the Greek philosophers. The hand of tradition, however, was heavy on them still. They dreaded to run counter to authority, and did not dare think unrestrainedly. Descartes shows us how we can understand things better if we will imagine a few principles by which it will be easy to account for things as they are. Then he carefully elaborates these principles as they occur to him; but he has no sooner done so than he takes care to add, "Of course, we know the earth was not made in this way."

A little later the philosopher, Leibnitz, believed in an orderly creation that had advanced by regular degrees, and that the lower animals had thus developed into the higher. He adds interestingly that there are probably on some other planets animals midway between the ape and man, but that nature has kindly removed such animals from the earth in order that man's superiority to the apes should be entirely beyond question.

By the middle of the eighteenth century men had begun to think more fearlessly. The great Emanuel Kant wrote in his younger and less timid years, "The General History of Nature and Theory of the Heavens." The great Newton had by his law of gravitation brought order into the heavens. Kant looked longingly for a greater Newton, who should find a similar unity in the animal world. He saw the wonderful likenesses between animals that the anatomist, Buffon, had recently pointed out. He believed there must somehow be blood relationship between all animals. He tried hard to conceive of some underlying natural cause by which all could have come about. As he grew older and his mind became more cautious he came to think the matter deeper than the human mind could ever fathom. He gave up the hope and believed the problem of animal origin and derivation would forever remain insoluble. He feared there was not in man the power to conceive his own origin.

If we ever wonder why it took so long before the thought of evolution should have fully dawned upon the world, the answer is not far to seek. No student of Natural History in ancient or medieval times had the faintest conception of the enormous number of animals and of plants in the world. The old Greek or Roman student of Natural History gives no evidence of knowing more than a few hundred animals. Men have named to-day, with systematic Latin names, hundreds of animals for every one that Pliny ever knew, and he knew more than any other man of early times of whom record has come to us.

In early days men who traveled into foreign countries brought back accounts of what they saw. The whole Natural History of ancient times was filled with the most absurd and ludicrous stories of all sorts of things to be seen in distant lands. Sir John Mandeville tells tales almost as imaginative and quite as amusing as those attributed to Baron Munchausen.

Upon the great awakening of the fifteenth century, with its new study and its wide-ranging travel, an entire change came over the human mind. Men who journeyed into far countries brought back with them not only accounts of what they saw, but, so far as might be, the things themselves. Collections of plants and of such parts of animals as could be readily preserved soon began to accumulate in every great center of Europe. It was only a question of time when such acquisitions must be arranged and classified, but as yet there was no system by which this could be done. The great Swedish botanist, Linneus, who lived in the eighteenth century, first taught us to give to each animal and plant two Latin names, the first of these to be the name of the group, known as a genus, to which it belongs, the second to be the name of that sort, or species, of animal. The cat, for instance, is Felis catus, the lion Felis leo, the tiger Felis tigris, and so on. Linneus then arranged the genera (plural of genus) into families, and these families into orders and so classified the animal and plant world as far as he knew it. In his earlier years Linneus thought of each species as being utterly apart and distant from any other. He believed it had been so from the first, each species having sprung in its complete form from the creative hand of God. In later life he came to show some evidence of the belief in development, but his great work is all built on the idea of the entire fixity of species.

About this time we find in the writings of Buffon, the French naturalist, many indications of an idea approaching our modern conceptions of evolution. He felt sure the pig could not have been a special creation, because he had four toes, two of which, with all their bones and their hoofs, are quite useless to him. We now call these toes "vestigial," and know the pig's ancestors used them, walking on four toes and not on two, as at present. Buffon believed there were degenerations as well as developments, and considered the ape a degenerate man. He conceived these changes to be brought about by what he called the favors and disfavors of nature. He varied much in his opinions in various parts of his career and occasionally is smitten either with conscience or with fear of authority. Then he goes back and says it is all a mistake and each animal is the product of a special act on the part of the Creator.

A little later, in England, Erasmus Darwin, the grandfather of Charles Darwin, who was subsequently to establish the evolution theory, wrote a long and elaborate poem called the "Temple of Nature." In this we find a remarkable prevision of many of the principles which were afterward to be warmly advocated and disputed during the growth of the idea of evolution.

"Hence without parents by spontaneous growth, Rise the first specks of animated life.

* * * * * *

Thus as successive generations bloom New powers acquire and larger limbs assume."

Erasmus Darwin recognized the struggle for existence, but he saw in it only a check against overcrowding, and not an active factor in the development as his grandson Charles came to see it. It is possible the elder Darwin's views might have been taken more seriously had he not clothed them with the form of verse. In these days it seems quite ludicrous to think of giving to the world a new scientific concept or a new phase of philosophy in verse.

The beginning of the nineteenth century gives us the first really great contribution to the idea of evolution. Under more favorable surroundings, this idea would have budded and become the parent stock of our modern theories. The chill frosts of adverse criticism by those in authority in science nipped the budding idea and so set it back that only of late years have men come to realize its strength and power. The Chevalier de Lamarck, serving in Monaco, was attracted by its rich flora to the study of botany. Coming later to Paris, he became acquainted with Buffon and was led by him to publish a Flora of France, using the Linnean system of classification. He was appointed to the chair of zoology in the Jardin des Plantes, and was given especial charge of the invertebrate animals, comprising all the members of the animal kingdom except those with backbones. After seventeen years of work over these forms, during which he wrote several books describing them, he finally published the great work on which his fame depends. This was the "Philosophie Zoologique." In this treatise he taught that the animal kingdom is a unit and that all its members are blood relations; that the members vary with varying conditions; that this variation results in continued advance. In all of these points Lamarck is at one with modern thought. His idea of the method by which the variation comes about has been accepted and rejected; modified, reaccepted, and again rejected.

Lamarck's conception of the cause of progress was somewhat as follows: The desire for any action on the part of an animal leads to efforts to accomplish that desire. From these efforts came gradually the organ and its accompanying powers. With every exercise of these powers the organ and its corresponding function became better developed. Every gain either in function or in organ was transmitted to those of the next generation, who were thus enabled to start where their parents left off. The general environment constantly gave the stimuli that led to the adaptive changes.

American zoologists have been especially inclined toward Lamarck's ideas. Until Weissmann startled the scientific world with his sharp denial of the possibility of transmitting to offspring any growth acquired by the parents, all seemed well. There is a tendency now to insist once more that slowly and gradually, in some perhaps as yet unexplained way, external factors do influence even egg cells, and gradually acquired characters do reappear in the offspring.

The blighting setback these views suffered came from the criticisms of Baron Cuvier. This genuinely remarkable man had built up the study of comparative anatomy. To him students flocked from all sides. Among these one of the most brilliant was Agassiz, the Swiss naturalist, who later came to this country, filled with Cuvier's ideas. This great teacher believed that species are fixed. He knew better than any man of his times the wonderful similarity in structure between animals of a given class. He attributed this not to any real blood relationship between the animals. They were alike because they had been made by the same Creator. This great Artificer worked along four main lines, and hence animals could be divided into four groups. Many who have studied text books on zoology written in this country by Agassiz and his followers will remember the four classes--Radiates, Articulates, Mollusks, and Vertebrates. Agassiz was such a wonderful teacher and so genial and so lovable a man that his opposition to evolution held back the advance of the Darwinian idea in America as Cuvier's influence had held back the Lamarckian idea in Europe. For the brilliant Cuvier simply laughed before his students at each "new folly" of Buffon and of Lamarck. Under this ridicule the influence of both men withered and died.

A little later the great poet, Goethe, turned his attention to the problem of evolution, giving an interesting account of the metamorphoses of plants. He declared, also, that the human skull is a continuation of the backbones of the neck, and that these bones have been transformed into the present skull. But his great genius as a poet drew his attention into other fields. Haeckel points out that if Goethe had known Lamarck's work his genius would have gained for the "Philosophie Zoologique" the interest and respect of the reading world. But Cuvier laughed it out of court, and only in comparatively modern times, since Darwin's work has set the world thinking anew, is Lamarck's career recognized at its true value. Lamarck should have been the founder of the evolution theory. But the time was not quite ripe, and it remained for Charles Darwin to announce his idea, sustained and fortified by years of careful observation and thoughtful reflection.

CHAPTER II

DARWIN AND WALLACE

We have seen in the last chapter that whenever men have actively thought they have attempted to explain the origin of plants and animals as well as of themselves. No one who wrote previous to the time of Charles Darwin had expressed any idea concerning this matter with force enough to convince any large portion of the thinking world. If Lamarck had fallen on better times, if the great Cuvier had not laughed him to scorn, if Goethe had found him out and made him known to the world, evolution might have come into its own sooner. None of these conditions arose, and it remained for Charles Darwin to give to the world in clear and cogent form the thought of evolution. He gathered so much material before he expressed his opinions, and looked at the matter from so many sides that, when he published his results, he had foreseen most of the objections which were subsequently to arise in opposition to his announcement. Charles Darwin is recognized to-day as the father of the evolutionary movement.

It has been sometimes said in recent years that Darwinism is dead, and there is a sense in which this is true. Unmodified and unassisted natural selection is not to-day considered by most scientists a sufficient agent for producing evolution. But everyone connected with the subject acknowledges Darwin as the master, and says that it was his work which converted the world to a belief in evolution. We can have no better preparation for an intelligent understanding of this subject than to consider carefully the life of this remarkable man and the circumstances under which he came to his epoch-making conclusions.

Evolution has taught us to attempt as far as may be to account for man on the basis of his heredity or of his environment. It is interesting to note that both of these factors in Darwin's case were entirely favorable. In the latter part of the eighteenth century Erasmus Darwin had given to the world an astonishing poem in which he anticipated not a little of the thought which his more famous grandson was to make so widely known. Josiah Wedgwood had learned to make for England her most famous pottery, no quality of which was more widely recognized than the sterling patience with which it was made. Erasmus Darwin, with his scientific proclivities, and Josiah Wedgwood, with his sturdy common sense and patient workmanship, united to give Charles Darwin his inherited tastes, for he was a grandson of both. Born in 1809, on the banks of the Severn in England, Charles Darwin was the delicate son of a practicing physician of modest but sufficient means. Owing to his lack of early vigor, Darwin spent much time in the open air, and in his excursions about his home was chiefly interested in collecting beetles. This taste, which lasted through all his young manhood, is the one early indication of the traits that were later to develop. At first in the day-school and later in the preparatory school Charles Darwin was anything but a satisfactory student. Even a kindly desire later to make the most of him makes it impossible to find traces of any especial fondness for earnest study. He himself believed his education to have been nearly useless, although he doubtless under-estimated its value. At the age of sixteen he went to Edinburgh at his father's desire, to study medicine. The sight of the dissecting-room nauseated him completely, and he refused to continue working in it. Later an operation which he witnessed in a clinic at the hospital sickened him so thoroughly that he declined to attend further operations. It became evident that the young man was not adapted to the life of a physician. The next move was to educate him for the church, and for this purpose, at the age of nineteen, he went to Cambridge. Here it soon appeared that he was no better adapted to the ministry than he was to the practice of medicine, and his university career went on in very desultory fashion. Most of his work was distinctly neglected, but two of the men he met there were to influence largely his future life. Henslow, the botanist, was unusually fond, for a professor in those days, of work in the field. Charles Darwin's tastes coincided with those of Henslow, with whom he formed an intimate friendship. He was always welcomed as a companion on the field trips. Though he studied little of botany in the classroom or laboratory, he was constantly with Henslow or with Sedgwick in the field. Sedgwick was the professor of geology, and of him Darwin was particularly fond, and under him did much the largest amount of his study. When he came up for graduation he ranked tenth of those who "did not go in for honors," a not very remarkable class standing. He was still required to put in two years of residence, and during this interval he spent most of his time with Sedgwick in the study of geology in the field. Returning to his home after a geological trip into Wales, Darwin found awaiting him a letter from Henslow, offering him an appointment that opened to his ardent mind the door to a career after his own heart.

The British nation, being the greatest commercial nation of the globe, has the greatest need for accurate charts of all the seas. Frequently she has sent out great charting expeditions to various parts of the world. One of these was to go out in Her Majesty's ship, Beagle, for a voyage around the world. Captain Fitzroy was in command, and he was especially commissioned to map the coast of South America from La Plata to Cape Horn and up the western side. In addition to this work, by carrying a set of accurate chronometers, he was to check up the longitude of the various ports to be visited in this circumnavigation of the globe. It was customary on such expeditions to carry a young man whose duty it was to study the natural history of the countries visited on the trip. The salary of such a naturalist was so small that an experienced man could scarcely afford to take the place. Therefore the appointment usually went to a man rather of promise than of achievement. Through Henslow's influence, Charles Darwin was offered this position in 1831. Darwin hastened to obtain his father's permission, but the elder Darwin at first declined to consider the matter. He felt that his son had not made such use of his time at the university as warranted the hope that much could be expected of such a journey. He believed it necessary that Charles should have some means of earning an adequate living before he could think of devoting his time to science. Charles found an efficient advocate in the person of his uncle, Josiah Wedgwood, Jr. Together they persuaded the father of the propriety of giving to Charles this opportunity to follow out his real tastes and ambitions. Accordingly, at the age of twenty-two, we find him embarked on a journey around the world. In the cabin of the Beagle he had abundant time, in his long sail across the Atlantic, to read the two volumes of Lyell's "Elements of Geology," which Henslow had handed him, with the suggestion that he read it, but on no account believe it. Filled with the love of geology as Darwin was, this epoch-making book was exactly the stimulus needed. Lyell had just begun to persuade the world that to understand the past we must study the present. In the forces now at work he saw cause enough to account for all the history of the past of the earth.

There is little doubt that this book was one of the most potent factors in determining the bent of Darwin's mind. His entire educational experience had failed to appeal to him. It is fortunate, we now know, that this was the case. If the university course of the time had really seized him it would have made but one more student like hundreds it was turning out each year. For most of us this is the happy event. Now and then comes the rare spirit to whom all of this fails to appeal because he is ready for something better. Such was the spirit of Charles Darwin. He started on his journey with a mind singularly free from prepossessions. In the long hours of this sailing voyage across the Atlantic Ocean Darwin had time to read and ponder Lyell's weighty words. By the time he reached the Brazilian shore he was filled with Lyell's conception that the present is the child of the past, developing out of it in orderly sequence. Lyell expressly denied that this is true of the animal and plant world. He applied it only to the face of the earth, with its mountains of uplift and its valleys of erosion. But the underlying principle of an orderly development under the action of natural causes was there. In Darwin's mind this at once found acceptance, and was destined to a fruition its author had expressly disclaimed.

The narrative of this voyage, as subsequently written, describes the islands visited by the Beagle in crossing the Atlantic Ocean. The contrast between the simple and general interest in these islands and the care with which Darwin described the Galapagos and the Keeling Atoll visited later in the voyage are speaking evidence of the rapid development going on in the mind of the young naturalist.

Reaching the shore of South America, Darwin first turns to its geology. But before long the animal life attracts his attention. In the Brazilian forest Darwin had his first experience of the wealth of animal and plant life in the tropics, and, like all naturalists, he was very enthusiastic over it. Among the animals that particularly attracted his attention was the sloth, a peculiar creature climbing slowly about the trees, small of size and sluggish of habit. Another animal that interested him greatly was the little armadillo with its interesting habit of curling up in its plated skin.

Captain Fitzroy soon finished what work he was required to do in this neighborhood, and Darwin was called back to the Beagle to continue his voyage. When they arrived at the mouth of La Plata their most serious work began. Here there was much tedious charting for Fitzroy, and Darwin could now leave the vessel for a lengthy trip on shore. This was doubly welcome. Seasickness was nearly constant with Darwin while on this entire voyage and every opportunity to work on land was eagerly seized. This region, too, was rich in objects of interest and in strange people. While exploring the pampas, beyond Buenos Ayres, Darwin came across the skeletons of the great mammals some of which Cuvier had previously described. He studied these bones with much care, and recognized at once in the megatherium a great similarity in structure to the sloth he had seen in Brazil. The enormous skeletons of the glyptodons struck him also as strangely similar to that of the armadillo. One evening, seated alone in the broad expanse of the pampas, the idea suddenly swept over him, stimulated, of course, by his study of Lyell: "Can it be that the little armadillo and the sloth of to-day are the degenerate descendants of the enormous megatherium and glyptodon of the past?" But his mind was not yet ready to accept so bold an idea and he swept it aside.

The people of this wild neighborhood interested Darwin very greatly, and he describes them with care. In this connection a charming trait of Darwin's character comes beautifully in evidence. The absolute purity of his mind, his utter freedom from grossness, shows clearly in his account of the first really semi-civilized people he had ever seen.

A little later, while exploring Patagonia, Darwin noticed the terrace-like formation of that desolate country. A flat near the sea was succeeded by a rapid rise, then came another flat. Three of these terraces in succession stretch back toward the Andes. At the base of the high terraces Darwin found marine shells, largely similar to those of the ocean beach so many miles to the east. His study of Lyell led him to suspect at once that this portion of South America had been raised in successive stages out of the bed of the Pacific. When they passed around Cape Horn and up the western coast he hunted for similar beach marks on the sheer western face of the Andes, and found them without difficulty, confirming his idea of the recent rise of this end of the Andean chain.

The Beagle continued its voyage up the western coast of South America until it reached Peru. Once more the abundance of tropical life is under Darwin's eyes, but now it is the life of an entirely different section. The dry climate of Peru furnished him with an environment distinctly unlike that of the moist Brazilian forest. He collects now with avidity, gathering especially insects and birds. Then the ship turned its prow westward across the Pacific, only to stop five hundred miles out at the Galapagos Islands. This little group he studied intensely, collecting large numbers of insects and birds. He had not worked over his collection long before he realized that each island in the group had peculiarities which marked its animals from those of any other island. Whenever two islands were close together in the group the differences in their fauna were found to be comparatively slight. If, however, he examined the animals from two islands lying at opposite ends of the group, the differences were always considerably greater. There was, however, a strong general resemblance among them all and a distant though not so strong resemblance to the corresponding animals of the Peruvian coast. On leaving the Galapagos group, Charles Darwin writes in his diary the suggestive observation that this little group of rocky islands seems to be one of the greatest centers of creative activity. It was this interesting resemblance of the animals of these islands to each other and to those of the Peruvian coast that finally persuaded Darwin that they were all related and were all descended from those of Peru. For the rest of his life, with an intensity which increased with each year, Darwin persisted in a patient search for the possible agencies by which such change could have been brought about. The problem, however, was temporarily eclipsed by a pressing geological question aroused by his visit to the Keeling Atoll. Here his investigation of coral reef formation absolutely captivated him. In the case of most coral islands in the Pacific Ocean the reef exists as a circle of coral enclosing a lagoon of water. In the center of this lagoon stands commonly a rocky island. It is plain that this is the foundation on which the coral built. But, in the case of the Atoll, the coral ring was present and so was the internal lagoon, but there was no rocky island.

The key to the solution came with an interesting discovery. Darwin began to put down a grappling iron on the outer side of the reef and to drag up coral. The farther away from the reef he went the deeper was the water from whose bottom he pulled the coral. What at first puzzled him was the fact that so long as he dragged up his coral from depths of a hundred feet or less the coral was alive. Whenever he went to depths of much more than a hundred feet, his coral was always dead, though he was evidently pulling it from situations in which it had grown. Then Darwin remembered the rising Andes, lifting themselves out of the bed of the Pacific. Here was the correlated movement. The bottom of the ocean here was sinking. As it sank it dragged down the corals with it. But the descent was so slow that new corals could build on top of the others fast enough to keep the reef up to the surface of the water. At the rate of growth of coral, this would seem to mean that the bottom could be sinking at a rate of only a few feet a century. But while the reef could keep up to the surface, the rocky island must slowly sink. Darwin inferred that there must be a rocky summit within the lagoon, below the surface of the water. A little sounding soon discovered this island, and the verification of Darwin's theory of coral reef formation was at hand. The description of this Atoll and of his theory of its formation won for Darwin the esteem of geologists when he later presented it in book form.

The voyage was continued around the Cape of Good Hope. Pursuing the usual course of sailing vessels, the Beagle touched once more at Brazil, returning home to England in 1836, after an absence of five years. Charles Darwin himself believed this trip to have been both his education and his opportunity. He had started on it a rather careless and indifferent student. He returned from it the most painstaking and patient naturalist the world has ever known. His father, who had hardly consented to his going because he believed him not stable enough to be intrusted to his own devices for so long a period, was profoundly moved at the sight of him on his return. Believing in phrenology, as did many of the physicians of his time, his father turned to his mother and said, "Look at the shape of his head; it is quite altered"; which, translated into the language of to-day, would read, "How wonderfully the young man has developed."

A part of Charles Darwin's duty to the British Government was to write a narrative of the voyage, and this account of his trip upon the Beagle is one of the great classics of travel in the English language. It won the confidence and respect of a wide circle of readers. In his next book he published his observations made at the Keeling Atoll and announced his theory of the formation of coral islands. This was a distinctly scientific investigation, and it won such immediate favor among geologists as to increase materially the young man's reputation. No one man is ever widely enough acquainted with the animal world to classify all the specimens gathered on such an expedition. In accordance with custom, Darwin began distributing his collections among specialists. Each of these was to identify and describe, to name, if necessary, the kind of material he knew best. Among others, Darwin had a considerable collection of barnacles gathered from boats and wharves in all parts of the world. As he could find no one sufficiently acquainted with these creatures to classify them he decided reluctantly to work them up himself. For about eight years much of his spare time was given to this painfully exacting work. He expresses himself as fearing it was a waste of time. Few systematic workers will agree with him. He did his work so well that it has been unnecessary for anyone to do it again. In addition it gained him the esteem of a new circle of scientists and that a decidedly exclusive circle.

The publication of these books did much for Darwin. His narrative of the voyage gained the good will of cultured England in general. The book on coral reefs won the geologists. His "Manual of the Cirrhipedia" (as the barnacle book was called) secured the attention of systematic zoologists. The time was not far distant when he would need every aid possible toward gaining and keeping the regard of men; for he was to promulgate a theory that would arouse the bitterest opposition and the keenest scorn.

All the while Darwin was working on these books his mind was quietly busying itself with what he called the species question. The more he studied the material collected on his long tour, the more confident he became that the animals of the present are the altered descendants of the animals of the past. He tried patiently to work out every conceivable hypothesis to see whether he could account for the alteration. He felt quite sure animals changed, but how they changed, and why, he could not for a long time conceive. He knew that gardeners were constantly producing new varieties of plants, and that animals of various breeds were clearly the descendants of other and familiar varieties. Accordingly he began to study the methods of animal and plant breeders, to visit their farms, to open correspondence with them and read all their trade journals, to undertake experiments in the breeding of plants. The longer he worked the more confident he became of the reality of the change; but for a long time no glimmer of the cause by which it could be brought about came to his mind. In 1838 he came across a book by Malthus called "An Essay on Population," in which the author shows that, whereas man increases by a geometric ratio, he cannot hope to increase his food supply in more than an arithmetic ratio. That is, while the food might increase like the series 2-4-6-8-10, the population would increase like the series 2-4-8-16-32. On this basis it is only a question of time when the earth will be too full of people for it to be possible for the food to sustain them. Malthus added many observations and suggestions, but this is as much of the book as interests us in this connection. Here was the idea that suggested to Darwin his agency for producing the change of the animals of the past into those of the present.

The number of animals of any particular species remains practically the same. There may be a few more one year, and a few less another, but on the average, year by year, the number of toads, the number of blacksnakes, the number of field mice, remains sensibly the same. Sometimes the rise of man brings an end to the wild population, and so in the past animals have dropped out of the race. Yet in the long run and for a considerable time the number of any species is constant. But each animal produces offspring in quantities sufficient to far more than replace himself as he dies out. In other words, animals increase not by addition but by multiplication. Too many are born for all of them to live. What becomes of the great mass of them? The answer is they die; most of them die young. Only a few fortunate individuals, favored by being a little stronger, a little more cunning, a little more attractively colored than their mates, survive to carry on the race.

The skillful gardener, looking over his flowers, finds a plant of more than ordinary beauty and thrift of growth. When it comes to maturity he keeps its seeds separate from those of the rest and next year plants them by themselves. As they come up he weeds out all unthrifty plants, only allowing the strongest to come to maturity. As they break into bloom he plucks away all whose flowers do not come up to the high standard he has set for himself. After a while he has but a few plants left, but these are the thriftiest and bear the most beautiful flowers. Again he allows these to mature and selects the seed of the very finest. Next year the process is repeated. After a few generations, usually three if the man is skillful enough, he has a definite strain of flowers that will thereafter come true. This is the process of artificial selection as carried on by man.

Darwin saw that Nature is constantly carrying on a similar process. She produces seeds enough on almost any plant to clothe the world in a few years if all of them could fall into proper ground and thrive like their parents. A friend of mine found a mullein stalk that bore more than seven hundred seed pods and averaged more than nine hundred seeds to the pod, a total of more than six hundred and thirty thousand seeds. If each of these could find lodgment on a plot eighteen inches square, produce a similar number of seeds and plant them all, the result would be overwhelming. The fourth generation would cover land and sea, from pole to pole, one hundred layers deep. But there is no such danger. Year by year the mulleins hold their own and no more. Any particular field may have more or less, but in the long run the average for a district is about the same. Some of the seeds are poor and thin. These scarcely sprout. Others spring up into thin-skinned plants, and the first frost nips them. Still others lack the woolly coating in its finest abundance, and the browsing animals eat these. Others lack power to put out a wide-ranging root supply and the first drought kills these. Still others fail to send up a vigorous stem and the passing animal knocks them over and they die. Of the few that are still surviving, some produce such small and inconspicuous blossoms that the insects scarcely see them, and they go unfertilized. In the end only the aristocrats of the group are left, aristocrats in the best sense of the word. These are strong, thrifty, and beautiful, and are provided with every defense known to the mullein world. From these the mulleins of the next generation will spring. Again Nature will select the best of these, by a repetition of the same process. Thus year by year the stock is improved. Any new feature that is favorable helps its possessor to survive, and, if happily mated, will show itself after a while in the entire group. This, in brief, is the underlying idea of Natural Selection, as Darwin conceived it.

In 1842, at Lyell's suggestion, Darwin wrote a short sketch of his ideas which he, two years later, expanded into a somewhat larger account. The manuscript of these early views of the theory was completely lost and has only been recovered within the last few years. It was recently published under the editorship of Charles Darwin's son, Francis. It is astonishing to see how clearly the first short sketch states the underlying conception which all of Darwin's subsequent work amplifies. Hooker was constantly urging Darwin to write out his whole theory in the form of a book, and Darwin had begun to do so in 1856.

Meanwhile, down in the Moluccas, Alfred Russell Wallace had been lying sick of a fever contracted during his exploring expedition in that neighborhood. He had been studying the distribution of the animal life of the Malay Archipelago. Overcome by sickness, as he lay in bed, he began to think over a book which he had read not long before, "Malthus on Population." Wallace had been pondering on the question of the origin of the animals of the Malay Archipelago. He had not the faintest knowledge of what Darwin was doing, but was influenced, of course, like Darwin, by what he read in Malthus. Interesting to relate, he had come to exactly the same conclusions, writing his opinions in the form of an essay. By the strangest sort of coincidence, he sent this essay to Charles Darwin, asking him to read it, and, if he thought it was not altogether too foolish, to send it to Lyell for publication by the Linnean Society. Darwin read with utter astonishment this essay containing views so absolutely like those that had come to him from his own long series of observations and reflections. With uncommon magnanimity his first impulse was to withhold his own publication entirely, but to this Lyell and Hooker would not for a moment consent. They were determined that Darwin should give them his long series of notebooks as evidence of the independence of his work and that he present to the Linnean Society, simultaneously with Wallace's paper, one of his own upon the same subject. In this manly form both essays were read at the next meeting of the society. The joint papers provoked instant discussion and prompt opposition. The world at large scarcely admitted a possible doubt of the fixity of species. Men generally believed the idea to be absolutely irreconcilable with their religious faith. Any question of the fact that the species of to-day exist practically as they had been handed down to the earth in the beginning by the Creator himself seemed to most men a direct blow at religion. At this time a very large number of natural scientists were clergymen, hence the opposition had abundant and influential support. The storm grew fiercer and more widespread. The publication in 1859 of Darwin's great book on "The Origin of Species by Means of Natural Selection or the Preservation of Favored Races in the Struggle for Life" added fuel to the flame.

In 1860 the British Association met in Oxford, and Bishop Wilberforce, the retiring president, in accordance with the custom of the society, gave a summary of the advance of science, especially during the preceding year. Everyone knew perfectly that the bishop would deal with the species question, and that he would handle it severely. Darwin was prevented by his usual ill health from being present at this meeting, but Huxley was there to see that their side of the question received proper attention. The bishop made a lengthy address, in the major portion of which he brought forward entirely worthy objections to Darwin's theories. Toward its close his feelings overmastered him and he departed from his manuscript and unburdened his mind. The lack of stenographers in those days and the tenseness of the moment, which made everyone forget to take down what was said, make it impossible to tell exactly what happened. It seems that Bishop Wilberforce, appealing to the prejudices of his audience, said, in language that now seems ludicrous but then was terribly bitter: "However, any of us might be willing to consider ourselves descended from an ape upon his father's side, no one would so demean his mother's memory as to imagine that she could possibly have shared in this descent." Huxley, who had waited patiently for the close of the bishop's address, saw immediately the fatal mistake. Turning to his companion beside him, he said, "The Lord has delivered the Philistine into my hands," and, rising, he hurled back at the bishop the indignant reply, "I should far rather owe my origin to an ape than I would owe it to a man who would use great gifts to obscure the truth." The bishop had made the mistake, and the struggle was on. Year by year it raged. One by one the scientists, first of England, and then of Germany, took their stand by Darwin. Huxley in England and Haeckel in Germany were the foremost advocates of the Darwinian idea. Long and fiercely the battle raged; slowly and gradually men began to see that, instead of undermining religion, the idea of evolution uplifted creation and made it not a strange happening in the distant past, but a divine activity through all time. But the battle had by no means subsided when one day came the sad news that Darwin's heart, so long feeble, so serious a hindrance to his work, had beaten its last on April 19, 1882.

His own people wished to bury Darwin quietly at his home in Down, but Darwin now belonged to the nation. A petition signed by many public men was sent to the Dean of Westminster, asking that his body might be granted burial in the Abbey. Probably no greater honor can come to man to-day, and fortunately Dean Bradbury was broad-minded enough to acquiesce. So it came to pass that the church that had so long believed him her enemy, that had first so bitterly fought him, came at length to see that he added a new dignity and worth to her faith, and took him to her bosom. Darwin's body lies buried in the Abbey.

In all the glorious company of immortal dead whose earthly frames are gathered in England's great mausoleum, there is no other one who has done so much to modify the mind of thinking man.

CHAPTER III

THE UNDERLYING IDEA

We have seen in the preceding chapters how the idea of evolution worked its way through the minds of men. Man after man got a glimpse of the idea, even among the ancient philosophers. But no one could speak convincingly on the subject before modern times, when a wider acquaintance with the animal world gave a body of facts on which it was safe to base conclusions. Even then the idea eluded men, until there came a worker trained by a long voyage around the world in which he had nothing to do except to study nature. He finally gathered in his mind material sufficient to convince himself not only of the truth of evolution but of the process by which this evolution was brought about. Every scientific principle is simple in its basal idea. In actual life the action of the principle may be so bound up with others as to need a skillful mind for its detection. But under all the complexities and modifications, like a silver thread woven into a cloth, runs the basal idea. Until a master has detected it the presence of it may be unsuspected. But once discovered and expounded, thereafter anyone may follow out its workings. So it is with the Darwinian idea of selection. It waited long for a discoverer, but, once found, we cannot but wonder why men did not see it earlier, it is so simple.

Mr. Darwin's mind, while slow and cautious, had a wonderful perseverance. When he had finished his work he had not only given a clear account of the process of evolution, but he had foreseen almost all the valid objections that were afterward to be brought against his theory. Some of them he had explained quite fully; of others he indicated a possible explanation; of still other questions he confessed that as yet they were not plain. But the whole theory is so simple in its fundamental ideas that it has completely revolutionized the whole aspect of modern biology and, indeed, of modern thinking in many lines.

There are four underlying conceptions, each simple in itself, which must be clearly perceived before one can understand Mr. Darwin's theory of "Natural Selection." The first of these is known under the name of Heredity. It is a matter of common observation that every animal or plant produces offspring after its own kind. Under no conditions would we expect a duck to lay an egg from which could hatch anything but a duck. No Plymouth Rock chicken mated with another of her own kind will ever lay an egg that will produce a Rhode Island Red. We may believe that the dog has descended from some form of wolf, but it is not meant that at any particular time in the past any wolf mated with a wolf ever produced pups that were anything but wolves.

Why this should be so is one of the most profound problems of biology. Nothing but the fact that the process has gone on under our eyes for so long a time could blind us to its marvelous character. To open the egg of a chicken and examine it by the most refined methods known to science is to find in it absolutely nothing that could be by the widest stretch of the imagination considered anything like a chicken. The biologist who has examined such eggs before and knows them in all stages of the process may recognize in an egg which had been incubated for a short time something which his previous experience tells him will become a chicken. But it has not the faintest resemblance to a chicken until later in its development. In early spring one may gather pond snails from any country stream and place them in an aquarium. The change from the cold water on the outside to the warmer water of the aquarium and the temperate climate of the room hastens the process which in the stream would not take place until later. In a short time one may find fastened to the glass side of the aquarium the little mass of transparent jelly which surrounds and protects the delicate eggs of these creatures. Fastened as they are it is easy to direct a magnifying glass so as to observe the change which goes on within these transparent eggs. It is even possible to apply a microscope in such a way as to watch the transformation under the low power of the glass. At first the eggs are as clear as water, having at the center a slightly yellowish spot. This central mass divides and subdivides until the separated sections grow so small and numerous as to lose individuality. Then the mass begins to press out here and dent in there. After a little while a double line of fine, hairlike projections runs around the creature. These hairs wave in such fashion as to make the embryo snail revolve slowly in its egg. A little later and swellings become more pronounced over the surface. One side flattens; the rotary motion stops; eyes appear at the front of the animal; a hump on the back begins to be covered with a shell, and the little creatures, pushing from the jelly, start their life journey on the side of the aquarium. Why did it happen? How did it happen? Here we have seen creation at work. Here surely the hand of the Creator is working in the only sense in which the Creator may be properly said to have a hand. How the history of the substance out of which the egg was produced provides for the future development of that egg no man has yet clearly said. This is not to say that we shall never know, still less is it to say that this can never be known. Ralph Waldo Emerson has said that there is no question propounded by the order of nature which the order of nature will not at some time solve. If he is right, and I believe he is, we shall at some time know how it is that this egg produces this snail. But, as I said before, nothing but the frequency with which the process goes on under our eyes could possibly blind us to the marvel of it.

The regularity with which each animal reproduces its kind is no more surprising than the faithfulness of that reproduction. Some of our birds have wonderful markings on their plumage. It is astonishing to see with what fidelity the feather of a bird may reproduce the corresponding feather of its parent. It will occur to everyone how, in the human family to which he belongs, there is some little peculiarity which, while not appearing in every member of the family, when it does appear is remarkably uniform. It may be only the droop of an eyelid, it may be a tendency to lift one side of the lip more than the other, it may be the peculiar shape of a certain tooth in the set, and yet when it appears it comes with astonishing similarity in all who possess it. So much for the principle of Heredity.

The second great underlying idea is known by the name of Variation. We have just been dwelling on the regularity with which parents produce offspring like themselves. We must now draw attention to the fact that, while it is true animals must absolutely belong to the same genus or species, even to the same variety, none the less no animal is exactly like his parents. Furthermore, in a group of animals produced at the same time from the same parent each one will have at least some small point in which he differs from every other one in the group. Two animals may look alike at first to the undiscerning eye, but a keen analysis of the measurements of the various parts of their bodies will show distinct differences. This is quite as true among lower animals. A toad may lay a double string of four hundred eggs which may be fertilized by the same male at the same time. These eggs may develop into tadpoles in the same pool not over a foot square. Within a few weeks these little toads may have gained their legs, lost their tails, and all may have left the water and taken to the ground upon the same day. Already the careful observer will notice differences among them. Some are larger than others, having grown more rapidly even though their surroundings were exactly the same; others are more skillful in their peculiar method of throwing the tongue at an insect they wish to catch. Still others will be differently colored. They might be arranged to show a considerable gradation between the lightest and the darkest of the group, though there may not be anywhere in the row a considerable gap. It is variation in animals of the same parentage and same surroundings which in the mind of Mr. Darwin made evolution possible. He always favored the idea that it was the continuous accumulation of these small variations that finally produced the profound changes which mark the new species. He admitted the possibility of the occasional appearance of those more distinct leaps in variation on which the present school of mutationists so strongly insists; but he believed them to be less influential, in the general trend of evolution, than the slower but much more frequent variations.

One of the most complicated and perplexing problems in the biology of to-day is the question of the origin of these variations. It is quite as hard to understand as is the method by which animals produce their own kind. No problem is being more earnestly studied. Suppositions we have in considerable number, and two of these at least may reasonably be mentioned. We will consider first the less certain theory. There is nothing in the egg that in the remotest degree resembles its parent. The old idea that every acorn had in it a miniature oak which only needed to unfold itself, or that the hen's egg had within it a miniature chick which only needed the warming process in order to make it evident, could not possibly survive the invention of the microscope. We may not, and we certainly do not, know everything that is in one of these eggs, but we do know most certainly that what is there has no resemblance to what it will be in time. The biologist finds in the nucleus or central core of every growing and reproducing cell certain minute bodies which Weissmann believes do much to determine the growth of the rest of the cell. He believes also that there are many more such "determinants" than are necessary for the reproduction of the cell. Each of these determinants may be fitted to produce slightly different results, but what decides which of them shall have its own way is quite uncertain. It may be that one determinant happens to be more favorably placed than others in the cell and that it has consequently secured more of the nourishment that comes to the cell in the blood of its parent. If this is true it would certainly be favored in the competition. We are becoming quite certain that whatever variations arise really start in the egg. The simplest conception as to the cause of variation would seem to be varied experience. One man trains his brain, another his hand; and in each case the organ so trained develops. But science is strongly of the mind that such influence does not reach the next generation.

A musician may have taught his fingers to be nimble; may have given them speed of motion and precision in their action. No child of his born after he acquired this wonderful facility of execution is any more likely to be a skilled musician than a child born before he had ever practiced enough to be anything more than a crude performer. Science is nearly certain that his children are just as likely to be talented along musical lines if he himself never had become a musician, simply because he had it in him to be a musician. In other words, they may inherit the talent which he developed, but they inherited it not because he developed it, but because it was in him to be developed. This is in accordance with the famous principle that there is no inheritance of acquired characters. We shall touch this question a little more fully in a later chapter, in speaking of the development of the evolution theory since Darwin's time.

If we are right in this matter, and we certainly are nearly right, variation must take place for the most part in the germ. These variations may not show until the animal has grown up, but they must have taken place among the determinants in the germ cell or they would not reappear in subsequent generations.

There is another process by which new variations may arise and which is more easily understood. It is the method of double parentage. The Barred Plymouth Rock chicken had its origin in such a double ancestry. The one parent was a Black Java whose color has disappeared entirely in the cross, but whose single comb with its few large points comes out clearly in the newly produced fowl. The other parent was a Barred Dominique. It is to this parent that the Plymouth Rock owes the interesting cross markings on its feathers. The comb on the head of the Barred Dominique is of the type known as the rose-comb, having many rows of slight projections. This has completely disappeared from the Plymouth Rock fowls. I am told that the skilled chicken fancier can tell, concerning many points in this fowl, to which of the crossed ancestors each quality is due. To a certain extent it is undoubtedly true that here we have the secret of the origin of many of those interesting people whom we are pleased to call geniuses. They may not possess any qualities not clearly discernible in various of their near ancestors, but in them we find what we, for the lack of a better understanding, call chance combination in one individual of the finer qualities of many ancestors, and this individual is so placed in life as to have these qualities developed and strengthened.

Charles Darwin, humanly speaking, may be accounted for as the happy combination of a double heredity and a favorable environment. He inherited the scientific inclinations of his grandfather, Erasmus Darwin, and the patient, sturdy honesty of his other grandfather, Josiah Wedgwood. These developed under the stimulus of the long five-year voyage, face to face with the world of nature. This happy complex produced the master biologist. To believe that he came about purely by chance requires a great stretch of the imagination. "There's a divinity that shapes our ends."

We have endeavored to make clear two of the basal ideas underlying evolution. One of these is responsible for the continued production of animals or plants of the same kind, preventing the world from becoming a wild kaleidoscopic and fantastic dream. Heredity is the conservative force of nature. The other idea underlies the development of new departures which keep the world from being a dull, dead, unending repetition of the same monotonous material. Variation is the progressive tendency in nature.

The third basal idea is that of Multiplication. Animals and plants multiply; they do not simply increase, they increase in a geometrical ratio. Anyone who has worked out one of these geometrical ratios knows how wondrously they mount up. There is an old familiar story of the blacksmith who asked the price at which the stranger would sell the horse he was shoeing. The owner of the horse replied that, if the blacksmith would give him one penny for the first nail he drove into the shoe, two for the second, four for the third, and so on, he might have the horse. No hundred horses in the world taken together have ever brought such a price as the blacksmith would have had to pay for the animal on which he was working. This is no circumstance to the awful story of what would happen to the earth if any animal could multiply unrestricted. The usual number of eggs laid by a mother robin for a single brood is four, and she may produce two broods in one season. This would mean that the original pair had produced eight offspring, four times their own number. If we can imagine these mating the next year and producing their kind in the same proportion; and, if we further suppose that each robin needs a space one hundred feet square from which to gather his food, we realize the astonishing fact that in fifteen years every patch one hundred feet square in Pennsylvania and New York would each have its resident robin, while the following season would find a robin on every similar patch from Maine to the Carolinas. Of course this could never happen, this is simply what would happen if all the robins could grow to maturity and reproduce at the normal ratio. But the robin is a comparatively slow producer.

Our turtles are more prolific. Twenty eggs would probably not be an unusual number. If we could imagine a turtle to live in the sea and to produce at this rate; and, if each turtle should need as much room each way as the robin, and a depth of water equal to its width, before the robins had spread over New York and Pennsylvania the turtles would have filled all the seas of the globe. Frogs are even more remarkable in this respect. Two hundred eggs is not an uncommon number. If each frog required a space twenty-five feet square on which to subsist, the entire earth would be more than covered with them within six years. It is ludicrous to think of such numbers, especially when we realize the hundreds of thousands of kinds of animals there are in the world, each of which is also multiplying, and it becomes evident at once that only an infinitely small proportion of all these creatures can possibly survive. This, then, is multiplication.

Here comes into play the fourth basal idea in Mr. Darwin's explanation. This is the part of Selection. When man produces new varieties of animals he does it by picking out from his flocks or his herds such as conform most nearly to his idea of what is desirable. These he mates, and from their progeny he selects the ones that suit him best. Generation by generation he gets his domesticated animals to conform more nearly to the standard of his desires. Natural selection works in exactly similar fashion. Of all the eggs that are produced by the animals at large in nature an overwhelming proportion never develop at all. They dry up, are eaten by their enemies, find no suitable place or time for development and decay, or are overtaken by some other calamity. Of the animals which emerge from the remainder an overwhelming majority come to an untimely end within the first few days of life. Each has countless enemies which prey upon him, and these have scarcely devoured him before they themselves become the prey of some stronger creature. Until Mr. Darwin gave us his elemental idea it was taken for granted that it was a matter of pure accident which survived and which yielded in the struggle and cares of life. It was Darwin who showed us that in this tremendous struggle against those of his own kind in the search for the same food, against the elements, in securing a mate, any animals possessing a superiority, however slight, must have some little advantage in the battle. Certainly, where so many must utterly fail, only those could possibly succeed who were well fitted to the circumstances in which they must live. We used to think animals were destroyed by the "accidents" of life and no one could foretell accidents. Mr. Darwin made clear that it was not a question of chance. That which might happen to any individual animal might be what we, not knowing the process, called accident, and yet there could be no possible doubt that those who succeeded were better fitted to battle with life than those who failed, and that their success was due primarily to their being thus advantaged. Consequently, if generation by generation the so-called accidents of life are constantly eliminating the unfit in overwhelming proportions, not only must the positively unfit disappear, but even the less fit. The more keen the struggle, the fewer could survive and the fitter they must be to survive at all. This is Selection. These, then, are Darwin's four great factors of evolution: Heredity, Variation, Multiplication, Selection.

From these it results that the animals and plants naturally become better adapted to the situation in which they are placed. When, as is constantly happening through the history of the earth, a change occurs in the physical geography of any region, when a plain is lifted to be a plateau, or a mountain chain is submerged until it becomes a row of small islands, this alteration will produce uncommon hardships among animals, even though they were well fitted to the old conditions. Any animal or any species of animals which meets such a calamity has before it only three possible outcomes of the struggle. First it may be plastic enough and it may vary enough in the right direction to adjust itself to the changed conditions. In this case it and a favored few like it will occupy the altered territory. The second possibility is that it may migrate while the actual change is going on, thus remaining in the sort of situation suited to it and its kind. The third possibility is the one which overtakes a great majority of animals--they die. Even the entire line dies out, and the strata of the rocks are filled with the bones, shells, and teeth of such as have met this fate. They have become extinct.

Thus far in this chapter we have been considering the influences under which it is conceivable that animals should advance. There is no question whatever that there are too many animals born, nor is there any possible question that a very large proportion of them must certainly die. There is equally no doubt that every animal produces after its own kind, and that its offspring, while they resemble it closely, still vary a little from it and from each other. This fact is perfectly plain to the most superficial observer who thinks on the matter at all. It is not so plain, nor is it easily demonstrated, that all of these acting together do surely, even if slowly, alter the form and behavior of the animal world. It is difficult to prove that there is going on under our eyes a steady and real improvement in the adaptation of the animals and plants around us to the situation in which they are placed. As far back as man's memory runs they seem to have been about what they now are; as far even as man's historical record runs they seem to have suffered no great alteration. The Egyptian of the old tombs is much like the Egyptian of the same rank to-day. The African of the tombs has the African features of to-day. Under such circumstances it is hard to prove that there is a steady and undoubted advance. For the most part the balance of the animal world is fairly even, and any species does not ordinarily change rapidly enough or migrate widely enough to show us its new features. It is difficult to see the struggle which we are so sure is going on. The life of animals is so hidden in many of its details that their joys and sorrows, if such we may call them, scarcely fall under our observation. Now and then an opportunity comes to see the process of adaptation work itself out. The struggle for existence begins anew and is carried on with special vigor, with victory, temporary or permanent, to one of the participants in the struggle.

The opportunity to observe such a change is presented in the United States by the introduction of the so-called English sparrow. This little creature, received at first with such joy, soon became the object of an almost bitter hatred on the part of very many people. This is really due to the fact that this bird is one of nature's darlings and thoroughly succeeds where it has an even chance.

The number of birds of any particular species which a region will support seems to be fairly definite. If a species is especially protected until it becomes unusually abundant, the removal of the protection commonly brings it down promptly to its original numbers. On the other hand, an accident of severe character or a special persecution may much diminish the number of the species, and still it will, within a comparatively few years, return to its previous abundance.

The inhabitants of Florida who own orange groves will never forget the winter of '94-5. A bitter cold wave swept along the coast and killed such large numbers of orange trees as almost to cut Florida out of the orange market and to open the gate to California, who was eagerly offering her fruit. This same frost caught the migrating blue birds and killed them by the thousands. When spring came bird-lovers throughout the eastern United States found an astonishing scarcity of these favorites. It was feared that with numbers so small they could not possibly compete with their enemies and with whatever untoward circumstances should be their lot. But there is room in this environment for a definite number of bluebirds. When this number was suddenly reduced the chances to make a bluebird's living were so wondrously multiplied that young bluebirds had such an opportunity in life as their fellows had not had for many long years. Accordingly they thrived as never before, and, of their progeny, a larger proportion lived to the following year. It was only a few years before the number of bluebirds had risen. Now we probably have as many as we have had for a long time past. I cite this simply to show that a region can support a certain number of animals of any one particular kind, and that the animal is likely to multiply, if given a fair chance, until it has reached such proportions. Now to my story of the rapid development of a newcomer.

In the year 1850 a resident of Brooklyn came home from a trip to Europe. He was a lover of birds, and while in Europe had been particularly attracted, no one now knows quite why, to the common House Sparrow, as it should be called. It is no more abundant in England than in many parts of the continent of Europe. A name that has been used for a long time is very hard to cast aside, and we shall probably continue to mistakenly call him the English Sparrow to the end. Our Brooklyn traveler brought home with him from

Europe eight of these interesting little birds and succeeded in inducing his colleagues in a scientific society to share his interest in them. Not wishing to commit the newcomers suddenly to the rigors of the American winter, these men built a large cage for the sparrows, meaning to set them free in the spring. For some reason or other when the winter was over the birds were all dead, and this first attempt to introduce the sparrow into America failed entirely. The little bird had won so many friends that his success was now sure. Finding a favorable opportunity, these Brooklyn men dispatched an order to a man in Europe, asking him to supply them with one hundred English sparrows. The consignment came in good shape and the birds were liberated on the edge of Brooklyn. This was the first of a number of introductions. A little later New York City sent for two hundred and twenty of these interesting creatures and turned them loose in her parks, while Rochester, with what was then considered great public spirit, purchased one hundred for herself. But the most progressive city in this respect was Philadelphia. She had long been troubled with the spanworm on her trees. This detestable larva had the unpleasant fashion of lowering itself by a long silken thread from the shade trees then so abundant in that beautiful city. The spanworms traveling around over the clothing of the passersby were so objectionable to everybody that it was with greatest delight that Philadelphia heard of the new birds which ate the pest. One wonders why some ornithologist did not look at the bird long enough to see that it had the sort of a bill characteristic of birds that eat seeds. It is true that most birds feed their young on insects, hence there is a time when any bird is apt to be insectivorous. But the structure of the sparrow's bill, like that of all finches, should have warned these bird-lovers that the sparrow was not to be depended upon to earn his living by catching worms. It is easy, however, to be wise after the event. Philadelphia believed she was engaging in a particularly advanced movement when she imported from England one thousand English sparrows, nearly as many as were liberated by all other cities together. These birds were turned loose among the shady streets and wide spreading parks of the City of Brotherly Love.

It is a serious matter lightly to disturb the balance of nature by the introduction of a new species. It is true that the sparrow did eat some spanworms and for a while enthusiastic bird-lovers hoped that here was the solution of the difficulty. Philadelphians will also remember that, with the spanworm removed from competition, the tussock moth, whose caterpillar carries on his back a series of yellow, red, and black paint brushes, at once become the permanent parasite of the long-suffering shade trees. This caterpillar is covered with bristling hairs, very closely set. Almost any bird objects to hair in his victuals; and this particular larva has hair more than ordinarily objectionable, for it irritates wherever it pricks the sensitive skin. This coating seems to protect the caterpillar from the sparrow, with the result that Philadelphia's trees were soon nearly defoliated by this comparatively new pest, worse than the spanworm. With the paving of the city's highways and the consequent shutting off of the air from the roots, the trees have largely disappeared from the streets of Philadelphia. With them have gone a fair portion of the tussock worms, but the sparrow holds his own. Here is a new bird in the field, and the struggle for existence on the part of every other kind of bird is now more complicated and severe. The sparrow can live where the rest of the birds have no possible chance. He throve so well in this country that by 1875 he had spread over five hundred square miles in the neighborhood of our larger Eastern cities. Thus far almost everybody was pleased with the new introduction. Within the next five years he had spread over more than fifteen thousand square miles, and wise men were beginning to feel doubtful of the virtues of their aforetime friend. When by 1885 more than five hundred thousand square miles had been occupied by the enterprising little fellow, there remained no longer a doubt in the minds of most people that the sparrow was an unmitigated nuisance and great fears were entertained that he had multiplied to such an extent as to be a serious menace. Here, then, is a modern instance under our own eyes of a victory in the struggle. If the sparrow has multiplied rapidly, while all the other birds have either only held their own or even have diminished in numbers, it is quite evident he must be better fitted to the conditions than they are. What are his fit points? Why does he succeed while others fail? The thoughtful bird-lover will have little trouble in understanding at least some of his victory-winning characteristics. How did he come to be almost the only bird who can live in large numbers in our great cities, without losing his ability to get along in less crowded situations?

In the first place this interesting bird is a clannish fellow. He has lost the ordinary sparrow habit and has come to like to live in crowded groups. Seclusion is not at all to his taste, and if there are only a few sparrows in the neighborhood those few will most certainly be found living near each other. One of the early adaptations of the sparrow to his city surroundings was the ability to find for himself a considerable proportion of his food in the undigested seed that could be picked up from the droppings of the horses. This naturally led the surplus sparrows out through the many thoroughfares leading from any large city. Where horses went sparrows could follow. Accordingly along the great lines of travel this bird found the simple path by which he could enter new territory. Meanwhile box-cars came into our large cities with freight. Sometimes they had carried grain, sometimes cattle. In either case it was not unlikely that a certain amount of grain should be found scattered over the floor of such cars. The sparrow visited these cars for the grain, and it must have been no infrequent accident that a door should be shut upon a group of sparrows, especially in inclement weather, when they were apt to be huddled in a dark corner of the car. These prisoners would be carried to the destination of the car and there liberated, thus producing a new center of what we are now inclined to call infestation. By such means the English sparrow has spread over much the larger portion of the American continent. Few birds are bold enough to visit a railroad car. Of the few who might be tempted, most are timid enough to fly on the first approach of man. Hence they fail to gain this chance of spreading. They must remain in the old crowded home. Meanwhile the sparrow, thus transported, finds a new home with fewer or no sparrows. The struggle is less keen. More of his kind can live. His boldness has been here a fit quality and has helped him in the race.

Man is only slowly coming to be a city-dwelling animal. Although it is a voluntary process with him, he still usually visits the country with much enjoyment. He has not as yet learned to adapt himself thoroughly to the city, for somehow city life kills him. Families that move into the city gradually have a smaller number of children in each generation until shortly the family is wiped out. The population of the city must constantly be replenished from the country. But the English sparrow is more adaptable than are the people. He has made himself at home in the heart of the biggest city. The Wall Street canyon is not deep enough, nor contracted enough, nor free enough of food to blot out the life of the English sparrow. At the heart of the deepest gully among the skyscrapers of our biggest cities we find this little bird hopping between the horses' feet, darting out from under the wheel of the push-cart, fluttering only a few yards to a place of safety, to return at once to his scanty meal upon the pavement as soon as opportunity offers. He is a typical city dweller and has learned to thrive there. Again in this matter he has distanced other birds to whom the city is more deadly than it is to people.

Another very important element in his fitness for the struggle of life lies in the fact that he is unafraid of man. He is wary of man; by which I mean he will quickly fly up from in front of man's feet. It is exceedingly difficult to catch a sparrow in one's hand. It is far easier to lure a pigeon within reach. But the sparrow, when escaping your hands, comes to rest but a slight distance away, only to elude you quite as successfully if you try again. If the sparrow is let severely alone he becomes more and more familiar with men, flies less promptly, and goes a shorter distance, but any attempt to trap him renders him shy more quickly than almost any other bird we have. He soon learns to avoid a trap in which his companions have come to grief. Those who would poison or trap sparrows must change constantly the base of their operations. This fearlessness of man is a valuable asset to the bird, for it is an important defense against other foes.

The most serious enemy the birds at large have, after man himself, is the bird of prey. Hawks and owls capture a large quantity of our smaller birds. Now the hawks and owls are for the most part shy of man. They have gotten a bad reputation, especially if they are of any size, because of their more or less pronounced proclivities for seizing our domestic poultry, and consequently many people will fire upon a hawk or an owl who would probably fire upon no other bird. By living close to man the sparrow is largely saved from the danger of capture by these carnivorous creatures, and this is the first and a very important element of the advantage to the sparrow of living near man. But there is the additional advantage that man scatters about him, in one way or another, a very considerable amount of waste food. I have suggested that the seeds in the droppings of the horse form a large proportion of the sparrow's food, and horses are to be found only with men. In the neighborhood of man's home, unless he has become sanitary to a degree which has only been attained in recent years, there is usually more or less garbage, kitchen offal of one sort or another. To this the sparrow has easy access and from it he makes many a meal. But this fearlessness of man gives him still another advantage which his competitors fear to use, it provides him with nesting sites.

Man has the faculty of putting up ornamental trimmings on his house, and there is no spot the sparrow chooses more willingly in which to build his nest than the ornamental quirks and cornices of man's architecture. A Corinthian column with comely leaves in its capital seems especially designed for the comfort of the sparrow, and his distinctly untidy nest is the familiar disfigurement of almost every ornate public building. These are the advantages which come to the sparrow from his willingness to associate with man, and there are comparatively few birds with whom he must share them. Few birds select the immediate neighborhood of man's home for their nests. They may live in the neighboring trees, they may haunt his orchard, but his house, for the most part, they decline to frequent.

Still another quality which makes for success in this buccaneer is the willingness with which he will vary his food as occasion requires. It is a not infrequent characteristic of the bird family that each species should have its own rather restricted diet. Birds are quite particular eaters, and many of them will come well nigh to starvation before they will use unaccustomed food. The sparrow, on the contrary, like man, eats almost anything he comes across that could reasonably be considered edible. He belongs to a group of birds which are structurally adapted to cracking the hard coats of seeds. This group of birds known as the finches is provided with the sort of bill familiar in the ordinary canary bird. It is short, heavy at the base, comes quickly to a point, and is firm and strong. With it the bird readily breaks through the hard outer coat of most seeds and feeds upon the rich cotyledons that are enclosed within. Nowhere in its entire structure does the plant crowd so much nourishment in so little space as it does in the seeds. It is not by chance that the great human food is grain. The sparrow belongs to the one bird group that makes a specialty of such seeds.

Most of the English sparrow's cousins in this finch group confine themselves rather rigidly to this diet. Here the variability of the sparrow again gives him the advantage. He may have the family fondness for seeds, but in their absence he can be content with almost anything edible. In the early springtime, when the seeds of last year are gone and those of the new year have not yet been produced, the sparrow is not averse to eating young buds from the trees. At this time he is not unlikely to eat our sprouting lettuce and peas. It is easy to be severe on him in this matter; but for a creature like man, who has the same tastes, who eats the enormous buds of the cabbage, the cauliflower, and the brussels sprouts, or the more tender buds which he calls heads of lettuce, it seems particularly inappropriate that he should throw stones at this little creature whose tastes are so similar to his own.

While seeds are more suitable for an elder bird they are altogether too indigestible to be the food of nestlings. So when the sparrow finds its nest full we know he must sally forth in search of nourishment more simple of digestion. Now for a few weeks he searches assiduously, catching insects and caterpillars of various kinds, and feeds them to his young. This taste passes as his children grow older, especially as shortly the seeds begin to ripen. Now is the time for the sparrow to fatten. Now he is eating the food for which he was really built. By the time the wheat is ripe there are sparrows enough about to form quite a flock, and when these settle down in a wheat, rye, or oats field and feed upon the grain, meanwhile shaking out upon the ground perhaps as much as they eat, the farmer begins to realize that the sparrow is not his friend.

When winter comes the struggle for existence among the birds is intensified, and comparatively few of them dare face it. Most of our birds betake themselves to less rigorous quarters, leaving to the sparrow a comparatively small number of competitors for the diminished supply of food. As long as the snow is off the ground the sparrows can find sufficient sustenance. They gather themselves into groups and sally out from the city into the open country. The immediate result is that great quantities of weed seeds are seized upon by the English sparrow, as, indeed, by every other finch which is with us in winter. Perhaps we have not given the little fellow credit for the good he does at this particular time, for the rest of the account truly does not help him in our esteem.

There is a further direct advantage in the sparrow's sociability. One robin may nest in the vines about your porch. If there were room for a dozen, scarcely more than one would be likely to use it, because he would drive away any other robin who attempted to share the neighborhood with him. To the sparrow company is always in order. While he may quarrel from morning until night with his fellow, it is a sociable quarrel and neither would willingly forgo it. This union is strength among birds, as with man. Every animal is safer from his enemies when he can have the constant presence of others of his own kind. The deer that stays in the herd is safer from the wolves. It is only when the latter succeed in cutting out some weaker or less sagacious animal that these carnivorous creatures succeed in tearing down their prey. I think the superiority of the sparrow over most of our common birds, when considered as a city dweller, is scarcely understood. Because he had won in the race with other birds is no necessary indication that he warred directly against them. Bird-men often attribute to him a quarrelsome disposition, as if he actually drove other birds away. It almost seems like animosity against the sparrow to speak of him as attacking blackbirds and crows. It is a cowardly crow who can be driven away by a sparrow, and if the two cannot live together it seems to me certainly to the discredit of the crow and not of the sparrow. I believe the truth to be that, while the sparrow is undoubtedly a quarrelsome fellow, his bickerings are his form of social converse with those of his own kind. A quarrel among themselves seems not to indicate animosity, but would appear to be the sparrow's idea of conviviality. It rarely leads to serious results. I have never seen a male sparrow trounce any other bird with half the vigor that I have occasionally seen the mother sparrow evince when she caught her male companion by the feathers of his head, hung him over the side of the limb, and vigorously and thoroughly shook him until he desisted from his annoying and possibly insulting attentions. The truth of the matter is that a colony of these little birds, with their continual social chatter, including their quarrels, makes such a continuous noise that the ordinary bird, which is generally of rather quiet disposition, is too much annoyed by the unending nuisance to find the neighborhood at all to his taste. Where a large number of sparrows have gathered together the conditions are such as would give a robin or a bluebird nervous prostration, and his only recourse is to depart to a neighborhood where there is more peace and quiet. But our English sparrow is not only better fitted for the struggle than the robins and bluebirds, the orioles and the wrens. He has one important advantage over even his own sparrow cousins. The males are handsome--much more so than the females or than their sparrow cousins in general.

In the song sparrow, field sparrow, chipping sparrow, and the fox sparrow the male and female are very nearly alike in color. It often becomes necessary for the bird-man to examine the internal organs of the bird he is stuffing before he can certainly decide its sex. But there is no difficulty whatever in telling the male from the female of the English sparrow. The male is far the more ornate bird. His back is striped with a richer brown; his head has two splendid dashes of chestnut over the eyes; his throat and breast are splashed with red and lustrous black; his bill is a clear fine black. Altogether the bird is strikingly colored for a sparrow, and this characteristic is the more remarkable when we see how quiet and somber is his more modest mate.

This brilliancy of male plumage in the presence of the somber color of his mate would seem to indicate that the English sparrow is eye-minded rather than ear-minded. It is true among human beings that most of them are eye-minded. That is to say, they notice things with their eyes chiefly. Memories they have are memories of things seen; recollections of their friends bring up the appearance of their friends. Their language is full of metaphors which imply form and shape. But occasionally we come across an ear-minded person. He remembers voices quite as well as he remembers faces. To him music is an unending delight, and painting and sculpture fall into a distinctly secondary place. This is ear-mindedness. Now, most of the sparrows seem to be ear-minded, at least as far as their recognition of their mates are concerned. In this group beauty of song is developed many times oftener than is especial ornateness of plumage. The bird-lover who is himself keen of ear is never tired of listening, when in the field, for the two low notes with which the vesper sparrow introduces a song, the rest of which is not at all unlike the one of his song-sparrow cousin. The field sparrow begins more like the song sparrow, but ends with an often repeated note, which not a little resembles in general character the somewhat more monotonous song of the grasshopper sparrow or of the chippy. In comparison with these melodious birds the English sparrow makes no showing whatever. His voice is harsh and querulous, although very occasionally it is possible for the bird-lover to detect a note or two which would indicate that, if he were properly educated, his voice might amount to something. He wins his wife not by his pleasant voice, but by his attractive appearance and his winning ways. We have every right to infer from the character of its fellow birds of the sparrow family that once the female and male sparrow were colored about alike. But Madam English Sparrow was apparently eye-minded rather than ear-minded. Whatever pleasant voice a suitor might have seems to have been to her without attraction, and there was nothing to encourage him in developing it, nor was she likely to mate with him for it and transmit it to her male children. On the other hand, let a suitor appear in whom a more brilliant coloring proclaimed his superior vigor, and this seems to catch her eye at once. The less accomplished rival in the tournament of love seems to have been already forgotten. To their children these successful characteristics were naturally handed on and led to equal success on their part. If any of these children possessed this badge of honor in a more than ordinary degree, he was the more likely to win a mate and thus again the opportunity of passing on to his offspring his own distinct advantage. Generation by generation the males have become more beautiful and the females more discriminating. That the bird is either instinctively or actually conscious of this advantage would appear from the constant fluffing of his feathers and spreading of his highly colored wings with which he evinces his admiration for his ladylove. Even the most hardened dweller in the city can scarcely have failed to see the sparrow spread his wings, fluff his feathers, and sink close to the ground, twirling and gyrating about the object of his affection. It must give him a shock to see how often she proves temporarily or hypocritically indifferent to the demonstrative proceedings. Indeed they may terminate in a thorough trouncing of the male on the part of the lady of his affections. Now this preference for color over song must have evidently evolved in connection with the development of social habits in the English sparrows. His cousins of the fields, our native sparrows, are much less social, much less likely to be met with in flocks. To birds who scatter more, beautiful song is a great advantage. It can be heard at a long distance. But when birds flock together a much better advantage is that of beautiful clothing, added to alluring ways.

But we have not nearly exhausted the catalogue of the traits belonging to our little friend which give him the advantage over other birds in the struggle for life. His ability to remain with us in winter when most birds are gone stands him in good stead.

It is readily observed by one who pays the least attention to outdoor life that winter finds us with comparatively few birds. North of Maryland and the Ohio River the robin is practically absent in the winter, except in much diminished numbers close to the border. The bluebird is similarly absent; the great flocks of blackbirds are gone; the bobolink is missing entirely; the thrush and the catbird have all left; the flicker and red-headed woodpecker are also spending their winter in the South. The great mass of our bird population has left us until warmer weather shall bring back to us once more our feathered friends. It is true that we are south to the snowbirds or juncos, and their little slate-colored bodies, with their light breasts and their white on each side of the tail, make our bare hedge rows brighter by their presence. A few of our birds like the song sparrow and the cardinal are hidden away in the thicket, and have not all joined their comrades in the south.

The English sparrow was once probably quite as migratory as any of the rest of these, but a great change has come over his habits. With his newly acquired fondness for the haunts of men he has suffered a change in this respect also. Whatever may have been his reason for migrating, it no longer holds. He now finds himself quite able to stand the cold of winter. Accordingly he never leaves us, except very temporarily. When the migrating season comes the sparrows of the neighborhood are very likely to gather themselves together in a single group and take to the neighboring country. I believe this flocking on their part at this time of the year is a remnant of the old migratory habit. Until snow covers the ground the sparrow is not likely to be seen again in such numbers in the city. The advantage the sparrow gains over his competitors by not going south does not appear during winter. When spring comes, however, his gain is evident. He has his choice of all the nesting sites in the region. When the migratory birds return every first-class place is filled by a sparrow's nest. Nothing but second choice situations remain, and with these the late comers must be content. When we consider how much the safety of the next generation depends upon the security of the young while helpless in the nest, we appreciate what the English sparrow has gained by staying throughout the year. Often while the season is so inclement that it would seem there is still danger of frost, the sparrow builds her nest. All sorts of places are open to her choice. She will find a protected corner under a roof, above a spout, in the corner of the porch, behind an open shutter, in the vines against the side of the house, on top of an old robin's nest in the tree, in the bird boxes which have been put up for more desirable creatures; anywhere and everywhere this industrious little mother is liable to build her nest. Her husband will help her more or less in the task, often bringing material and helping to place it in the negligent pile of which their nest is composed. But he does a good deal more fussing and cheering up than he does actual work, and she seems to depend much upon his cheerful presence for her happiness. It is hard to discourage Madam Sparrow when once she has set her mind on home-making. A bird-lover, some time since, reported how a pair of sparrows had started to build a nest upon his lawn. He, wishing to interfere with the process, took a small rifle and shot the male bird. Within twenty minutes the female, who had scouted round the neighborhood, returned with another mate and resumed her nest-building process. Again he interjected the tragic note into her life by shooting her second husband, only to find her start out in pursuit of a third, with whom she returned in the course of an hour. He felt that by this time he had interfered with her domestic happiness as much as he had any right to do, and suffered her to continue her housekeeping with her third husband without further molestation. I imagine it would have puzzled both birds to tell who was the father of the nestlings who appeared two weeks later.

Not only do sparrows nest early, they nest often. I suggested to one of my students that she locate as early in the season as she could the nest of a pair of English sparrows, which was sufficiently accessible, and that she keep it under observation at intervals of a few days throughout the summer. In the fall she came to me with glowing eyes and gave me her report. "It is simply great," she said. "I never went to that nest a single time this summer to find it empty. When I first got there I found four eggs; after a while these hatched out, and the young were on the nest until they were old enough to fly; but before they had left she had slipped a fresh egg among them, ready to start a new batch. Whenever I saw the nest throughout the entire summer, I found in it either eggs, or young, or both." Such reproductive energy as this is hard to beat; compared with this rate of increase, the ordinary bird is the exponent of race suicide. How can a robin hope to compete with this family industry? What can a bluebird offer that will approach such chances of a worthy successor when his work shall be finished?

These, then, are the most important points in which the English sparrow has varied from his sparrow cousins and made of himself the most successful town dweller in the bird world. He has become clannish and gained the advantages of cooperation. He has used man's highways and cars by means of which to expand his area. He has cultivated the presence of man and thus gained protection from his enemies, food from man's waste, and nesting sites on man's house. He has assumed a varied diet. The male has become handsome. He has given up migrating, and thus secured the best nesting sites. He has learned to produce many offspring. With all his versatility, why should he not succeed?

Thrown into competition with our native birds, he easily beats them on their own ground. He survives against the competition of birds which seem to us more estimable in every way. The very fact that he survives proclaims his superiority over them, and shows that our criterion is not the one by which nature judges. We like the birds which serve our purpose. We admire the brilliant plumage of the jay, cardinal and goldfinch. We love the mellow notes of the woodthrush, and of the veery, the clear, rollicking outpourings of the bobolink, the musical love song of the brown thrasher, the cheerful scolding of the wren. We are fond of the birds who busy themselves taking the insects out from among our grain and from off our fruit trees. We can only understand the value of the bird to nature when he is valuable to us. So, because the English sparrow does little that is to our advantage and much that is to our annoyance, he is in our estimation a reprobate and an unending nuisance.

All sensible bird-men must clearly acknowledge that he is a very undesirable citizen. I write the above sentence to show that I realize the whole duty of the bird-lover in the matter of the sparrow. This pestiferous creature should be exterminated by traps, by grain soaked in alcohol, or strychnia, by fair means or foul. But personally, I am taking no share in his destruction. Any bird-lover, after reading the foregoing account, can scarcely have missed the undercurrent of my affection for the little rascal. He is a thorough optimist; he is absolutely persistent; no hardship seems to dampen his ardor. His heart is valiant above that of most birds so that he has dared to make of man his near neighbor when other birds consider him their worst enemy. I love him for it. When I am in the midst of a big city with its cliffs of offices and its gorges of paved streets, it is to me a cheer and a delight to see this happy little fellow who has adapted himself to circumstances against which no other bird, excepting the pigeon, can cope. I confess that it would be with regret that I should see him disappear from the landscape. I have missed a long line of spring peas through his ravages, and he has objectionably decorated many places about my own home. But I have yet the first violent hand to lay upon the sparrow, and I doubt whether my hand is ever to be reddened with his blood.

I am going to ask bird-men to forgive me if I say that I believe, although I speak only from general impression, and not from careful research, that the sparrow within the past eight years has reached his equilibrium in the neighborhood of Philadelphia and is growing no more abundant. Meanwhile another and very desirable state of affairs is arising. Bird love and bird protection are so active in this neighborhood that there is growing to be a new race of birds who lack the fear of man their ancestors justly had. Under these conditions the wild birds, which for a while we believed to have been completely driven out by the sparrow, are rapidly returning to our villages and towns, and we have many more robins and catbirds, wrens and flickers than we had ten years ago. We have seen the worst of the English sparrow;

he has now found his equilibrium.

CHAPTER IV

ADAPTATION FOR THE INDIVIDUAL

Among the standard books of the classical curriculum in the denominational college of thirty years ago was a volume which I suppose has practically disappeared from such courses. It delighted many of its students for a reason entirely different from that which the author meant should be its taking feature. It was Paley's "Natural Theology." The author started with a story of a watch found by a savage. This child of nature was supposed to examine its mechanism and to infer that the watch was made for a definite purpose. As I remember, he was even supposed to discover that its purpose was to mark time. It was at least to become clear to his savage mind that this was no chance object, but was the definite product of a designing mind. Having brought this hypothetical savage to these conclusions, the author turned himself to savages nearer home who fail to see design in nature. The book takes up a great many cases of interesting facts in animals and plants as clearly showing evidences of design as did the watch our savage picked up. But the inference we were expected to draw was that the design shown in nature argued clearly for a Designer above nature; in other words, that nature was unintelligible without God. Everyone in the class believed in God without this preliminary, and consequently the book was unnecessary, so far as we were concerned. We started with the condition of mind which the author hoped to produce. One effect the book did have; in the absence of any other reputable course in zo鰈ogy, it gave us an astonishing collection of interesting facts about animals.

Some of Paley's statements were certainly peculiar. His Malay pig with its upper teeth wonderfully curved was said to be in the habit of hanging its head upon a bush while it slept, in order to save the strain upon its porcine neck. This was too much even for our credulity. None the less the impression made upon some of us by the evidence for design in nature has never left us.

Among many scientists to-day it is supposed to be crude to speak of purpose in nature, and there is reason for their attitude. But the statement that there is no such plan conveys to the ordinary thinker a meaning that is far more erroneous than could possibly exist in his mind should he believe implicitly in design and purpose. As between design in the universe in the usual sense of the word, and a purely accidental connection of events in the universe, there can be no doubt as to the choice. The truth is far better expressed by the word design than by the chaos which is the alternative idea in the average mind. In these later years we have come to use a different word. We now conjure in such connection with the word adaptation. In every animal and every plant the trained eye sees unending examples of adaptation; that is, of a fittedness to the work it has to do. The modern scientist feels sure not only that the animal is fitted to his work, but that he has been so fitted by the work; that the very use he makes of his organs has determined their structure. This work has decided that the structure which he has is the structure that shall survive and shall produce other structures like itself. Adaptation therefore does not simply express the idea that the animal is adjusted to its surroundings, but it further suggests that the animal by gradual process has become thus adjusted. The word adaptation applies not simply to the result, but also to the process. The scientist does not consider the animal a final and complete result. He thinks it still in a state of flux, and so long as its line lasts it will be in a state of flux. Change is about it on every side, and it must adapt itself to this change or it will pass away. It may adjust itself, as has been previously stated, by moving to another environment in which it feels more at home, but unless it does this, if there come much change in its present surroundings, it must either meet the difficulty by altering itself, or it must give up the struggle. The alteration is unconscious so far as the animal is concerned. It is seriously to be doubted whether there is any recognition of the process on the part of any animal excepting man. But though the process be unconscious, it is none the less there. Slowly and gradually the animal and the environment are becoming adjusted to each other.

While it is exceedingly difficult to lay our hands on any animal which is at present visibly changing its structure, it is not hard to find closely related animals. These are nearly alike in structure in most respects. In a few points, however, they may differ materially, and these points are often directly concerned with different habits of life. Considered in this aspect, these adaptations of a single organ separately examined form an excellent argument in favor of that gradual alteration of the entire organism which evolution suggests.

The most primitive struggle in which an animal can possibly engage is the effort to maintain its own life and vigor. This struggle will result in certain adaptations for the individual, adjustments which make for the safety of the animal himself. These form the subject matter of the present chapter.

The farther up the animal kingdom we pass in the study of adaptation, the more likely we are to find changes which have but little bearing on the safety of the individual. They work for the good of the entire species, sometimes to the distinct disadvantage of the individual. The King Salmon may make its long run to the headwaters of our western rivers, deposit its eggs, have them fertilized, and then float down to death. But it does not die before abundant preparation has been made for the continuance of the race. Such adaptation for the good of the species will be considered in the next chapter.

The first and most important struggle any animal has to enter is the never-ending battle for its food. Occasionally there is a similar straining after the air it breathes. But ordinarily air is sufficiently abundant, except to animals living in the water, where the supply is always more or less restricted and easily becomes exhausted. But food is the constant need of every organism, and most creatures die for lack of it. In this struggle the animal is pitted against those of his own kind, rather than against those of other species. Even his brother is his enemy, for he desires the same food. In many a nest of birdlings one of them fails to reach its development simply because the parent either is unable to find or it cannot carry enough food to satisfy all the hungry mouths in the same nest. Before the nestlings are ready to take their place in the struggle for life outside and hunt their own living, one or more of them has succumbed.

After the battle for food comes the struggle for shelter. For most animals there is no such thing as shelter. They are exposed to the inclemencies of the weather and to the depredations of their enemies without the means of retiring into any situation which might protect them. In the higher animals, especially when they are warmer blooded and their bodies must be kept at a higher temperature, some form of covering has come to be almost universal.

Though comparatively few animals are prepared to seek shelter from the cold, all of them have enemies against whom they must battle. These foes may wish to eat them or may simply wish to get them out of the way. In either event this struggle is so persistent and so keen that after starvation it is probably the source of the largest loss to the animal kingdom.

Considering first the feeding habits of animals, we find they are exceedingly varied. Some creatures simply engulf other and more minute animals, often only microscopic in size, in such quantities as to satisfy their hunger. Others, feeding upon larger plants or animals, must have some means of breaking off particles of this food; still others confine themselves entirely to nutritious fluids, and must have organs adapted to this particular type of food.

Insects are so common that anyone, who cares to, may easily verify what is here described. It will take nothing but a clear observant eye and a little patience to make out what is suggested. Each of our common insects has one of two clearly defined habits in the matter of food. Either it eats solid food, which must be made fine before it can be taken into the mouth, or it feeds upon liquids. These liquids may be easily accessible like the nectar of flowers, in which case one sort of mouth will serve; or they may be the juices inside the tissues of animals and plants, when an entirely different type of mouth must be employed in their acquisition. Perhaps the most easily found representative of the biting type of mouth, which breaks up solid food, will be seen in the common grasshopper. Doubtless each one of my readers has at some time taken a grasshopper into his hand, and, holding the tip of his finger against the insect's mouth, has promised the creature its freedom on condition that it disclosed its reprehensible habit of chewing tobacco. The grasshopper surely complied, and I trust the promiser was as good as his word. The grasshopper's head is so placed that, while it is at the front of its body, the mouth is directly on the under side of its head, while the eyes are at the top of the front of its face. Under these circumstances it cannot see what is going into its mouth, and this makes an interesting variation of conditions to which it must adapt itself. The means by which it accomplishes this will be clearer if the mouth of the grasshopper be compared with our own. Our lips are upper and lower, but the grasshopper has a front lip and a hind one. The broad front lip is easily seen at the forward side of the mouth. Just behind it, serving the purpose of our teeth, is a pair of hard jaws with horny tips upon them, which serve to break small pieces from its food. While our jaws and those of all other backboned animals work up and down, so that we may be said to have an upper and lower jaw, the grasshopper and all of his insect, crab, or spider relations, which have jaws at all, have them right and left, and they work from side to side. Behind these harder mouth parts is found a pair of softer jaws, each of which has on it a little finger-like feeler. With this pair the insect holds its food while the hard jaws break it to pieces. The hind lip follows, and is also provided with short finger-like feelers. The feelers on the hind lip and on the soft jaw are necessary because the eyes are so placed as not to be able to see what goes into the mouth, hence the insect must make up for the loss of sight by the addition of touch. The same type of mouth as the grasshopper has will be found among the beetles. Here the males sometimes have the hard jaws so enormously enlarged that they are known as pinchers and have given to their owners the name of pinching bugs. All insects with such jaws as these use them for breaking up solid food.

A glimpse at the mouth of the butterfly captured on an adjoining flower will show a most remarkable variation from that seen in the grasshopper. Practically all of the mouth parts mentioned are present in this insect, and its early ancestors had their organs practically like those of the grasshopper. Now they are so modified and united with each other as to be almost unrecognizable. The pair of soft jaws has become very much elongated, and they lock together in such a way as to enclose a hollow space between them through which the creature can suck its fluid food. Not only have these soft jaws joined together, but, because they have become so much elongated when not in use, they must be coiled up like a watch spring and laid between two hairy lip-like processes which correspond in reality to the two finger-like feelers of the grasshopper's hind lips.

The butterfly, lighting upon the corolla of the flower, uncurls this long "tongue," and through its hollow center pumps up into its crop the nectar which the flower has stored in its base. When the butterfly comes to get the nectar from the flower, it rubs upon its own hairy body pollen from the stamens of the flower and carries it to the pistil of the next flower of the same kind which it visits. Most of us have at some time sucked the nectar from the back of a torn honeysuckle blossom and approved the taste of the butterfly in this matter. If the airy creature be watched as it lights upon a flower, it will not be difficult to see it uncurl this long tongue and probe the depths of the flower. If the butterfly be taken in the hand and the tip of a pin inserted in the center of the coiled tongue, it can be uncoiled without the slightest harm to the butterfly.

Insects which wish to use for their food the juices of other animals or of plants do not find them so easy to gather. In the mosquito most of the mouth parts are developed into slender pointed bristles wrapped in a hind lip. These bristles serve to puncture the skin of the creature attacked, while the curled lip serves as a tube through which the blood may be extracted.

If, while sitting on the porch on a warm summer evening, mosquitoes begin to annoy, let one of them at least serve to show his method of procedure before he is destroyed. Allow the creature to alight upon the back of your hand and slowly raise the arm until the eye looking at near range can see the head of the mosquito, which, by the way, is sure to be a female. Males in this species are entirely harmless. They never eat after they have grown up; that is, after they are truly mosquitoes. But the female is very assiduous. Alternately raising and lowering her lancets from either side, she pierces, then saws, her way down through the flesh until she has buried her instruments in her victim and her head rests against her prey. Now a pumping motion of the abdomen will be apparent, and this continues its accordion-like action until it becomes more and more distended. The insect only gives up its task when the entire abdomen is swollen into a great red ball of blood. The mosquito will now slowly withdraw its instruments and retire from the scene, if permitted to do so. If there is any fear of annoyance from the bite, a drop of ammonia immediately applied will counteract any irritation which would have been produced by the saliva of the mosquito. The insect is not intentionally vicious in this procedure. It is simply gathering its own natural food, though this does not make it less annoying to us since we are its victims. The swelling produced after the bite is the result of the action of the saliva the mosquito injected into the wound. The opening through the tongue is so small that blood would readily clot inside the tube and prevent its further usefulness, did not the mosquito inject the secretion of its salivary glands into the wound. This acts upon the blood in such a way as to prevent its coagulation.

Anyone who thinks carefully can add numberless specializations for food getting. For instance, primitive mammals have little pointed teeth which fit them for feeding on insects. In each of the great order of mammals a special development of these teeth has occurred. Among the rodents or gnawing animals the front teeth have become long and chisel-shaped for nibbling. The horse has formed them for nipping, and his hind teeth for grinding. In the dog the teeth near the front have become long for tearing his flesh food, while his hind teeth, working with the motion of scissors, cut it into pieces.

A second great class of specialization is seen in the changes of habit that provide the animal with shelter. The home seems so necessary a part of human life that it is almost impossible to think of an animal having nothing that in the faintest degree could be called a home. We at least expect it to have some sheltered place in which it passes most of its time and to which it returns after its wanderings. The great majority of all animals have no such home. The place in which we find them to-day may not be the place in which they will be to-morrow. All places are alike to them. The ordinary conduct of their daily life drives them about in the search for food. Their attempt to escape from their enemies leads them each day into new situations, and they may, and probably do, have no power to recognize the old location if they return to it. When we come to the backboned animals there is a little more tendency to a stationary location. The sun fish may frequent the same reach of the stream, the trout may haunt the same pool, year after year, but a great majority of fishes doubtless move indiscriminately up and down the stream or about the lake or ocean and are not found two successive days in the same place. The same may be said of frogs. For a time a particular frog may have a fondness for a special bend in the stream, but it is only a temporary fondness, I believe.

Our own need for shelter is the prime motive in leading us to build a home, and this necessity arises first of all because of our warm blood. What we are accustomed to call cold-blooded animals are not truly so. Their blood holds practically the temperature of their surroundings. As the air or the water in which they live grows warmer or colder the bodies of these creatures alter with it. Consequently they are active when the temperature is high and grow more sluggish as the thermometer falls. When the day grows distinctly cold the animals may go practically dormant.

Only the birds and mammals have warm blood, and of these the birds are distinctly the warmer. Whereas the temperature of the mammals runs from about ninety-eight to a hundred degrees Fahrenheit, that of birds lies somewhere between one hundred and five degrees and a hundred and ten. Creatures which are warmer than their surroundings must have some protection against chilling. Accordingly both mammals and birds have clothing. In the case of mammals the covering is fur, in the case of birds feathers. In some of the tropical animals like the elephant and rhinoceros, or in man, who has learned to protect himself in cold regions by making clothing for himself, this hair is very short, and except where serving for ornament is quite scanty, no longer being of use as a protection. But the great majority of all mammals are well covered with a dense coat of hair. In many of those living in the colder regions there is in reality a double coat. The fur seal of the Alaskan Islands is so provided. A set of long hairs deeply fastened in the skin forms a covering, which shows on looking at the seal. Underneath this layer, and set but lightly into the skin, is a short coat of very much finer hair known as the underpelt. When the skin is taken from the seal it is split by machinery into a lower and an upper layer. When so split the deep-seated pits of the long hairs are cut, and these hairs come out. The fine underpelt thus laid bare is what is commonly known as sealskin. Fashion has decreed that this must be dyed a rich brown, although when taken from the animal it is nearly mouse gray.

The birds have need for better clothing. To begin with, their blood is much warmer, and hence needs better protection from outside cold. In addition such of them as fly high must be prepared to stand great variations in temperature. For these purposes birds need a covering of the finest type. This clothing, in addition, must be extremely light because the creature must carry it into the air in flight. All of the requisite conditions are thoroughly met by the feather, which is the lightest and warmest clothing known to man. If at night we wish, regardless of expense, to keep ourselves warm with the lightest and warmest of covering, we send to the Arctic Sea, and from the breast of the eider duck we pluck the down which lies between the warm blood of the duck with its temperature of one hundred and seven degrees and the water in which the iceberg floats.

Young mammals and birds, before their clothing has well formed, are naturally susceptible to cold; this leads to the first genuine approach to a home among animals lower than man. Birds lay their eggs long before the creatures inside of them are ready to emerge. Accordingly they have learned to build nests in which to place these eggs, and to protect them from the outside air; meanwhile the bird keeps the eggs warm by close contact with its own body. The lowest of the birds may lay their eggs simply on the ground without any special protection. As we rise in the scale of the bird world we find nests provided for the eggs. These nests become increasingly complex and specialized, until we reach the oriole's home with its wonderfully woven mass of fiber, which, in spite of its apparent looseness, supports well the weight of the mother bird and of her eggs. The robin, not content with making a woven basket, plasters it with clay, and makes an absolutely impervious nest.

When we remember that both mammals and birds are the modern descendants of cold and scaly reptiles of an earlier geological time, it becomes interesting to compare their clothing. Evidently in the mammals hairs began to come out between the scales. Gradually the scales became fewer and the hairs more abundant until finally the scales have all disappeared, except those that remain as the claws on the toes. The ancestors of the birds, on the other hand, boldly transformed their scales into feathers.

Another need for shelter arises in connection with the approach of winter. This problem of withstanding the cold season is complicated by the presence of two new factors. First and most directly, the cold itself is a distinct obstacle to the comfort of many of these creatures; as a secondary result of this cold, the food of many animals disappears entirely in winter. Most of our birds meet this difficulty by changing their base of operations. When the north grows cold these creatures fly to the south. Some of their migrations cover enormous stretches of country. Our bobolink, so well known and loved by all watchers of spring migrations, passes twice a year between the latitude of New York and Rio Janeiro. One of our most careful students of bird migration says that the Golden Plover makes, twice each year, the long journey from the Arctic shores of North America to the plains of La Plata.

Different fur-covered animals have specialized to meet the winter by any one of three different methods. They may brave it out, hunting for their food as best they can all winter long. Such a course is pursued by the rabbit. Again like the squirrel, they may store large quantities of food during the summer, and on this provender they may subsist during winter, remaining for most of the time near their hiding-places, which, however, they may frequently leave upon warm days. A third method is less common, but very interesting. The groundhog or woodchuck is the best-known example of the group. It remains asleep, or, as it is technically known, dormant, during the winter. This stupor is more profound than ordinary sleep, and from it these animals awaken with difficulty. It is needless to remark that the groundhog's behavior on the second of February has no relation whatever to the weather we are to have later in the season. This is coming to be pretty generally understood. While the newspapers each year comment upon the groundhog and his shadow upon that day, year by year the notice has more of humor in it, and fewer people pay any attention to it.

As for the backboned animals which are cold-blooded, these must, unless they are fish, give up the struggle completely, bury themselves in out-of-the-way places, and go worse than dormant. They often become absolutely cold and stiff. In the case at least of fish, it is quite possible for them to be frozen stiff, even to be enclosed in cakes of ice, and still to recover if the encasement is not too long continued. But the snakes, the turtles, the toads, the lizards, all are hidden beneath the ground waiting in absolutely unconscious rest the return of warmer weather.

After the need for food and shelter comes the continually recurring necessity on the part of almost every type of animal to escape from the unwearying persecution of higher creatures which would feed upon it. The whole creation is a constant network of animals which prey upon each other. It is the fate of a great majority of all creatures to fall victim to other animals to whom they serve as food. Accordingly nature has concocted many devices by which she assists her favored children in escaping this relentless persecution. Perhaps the most widespread means which animals have developed in order to elude their enemies lies in the possession of power to escape their attention. Two different factors may contribute to this end. The first of these consists in the practice on the part of many animals of remaining absolutely quiet in time of danger. This instinct seems to be nearly universal. The first impulse of most animals upon discovering danger is to remain absolutely motionless. The eye detects, with ease, objects in motion. These same objects might entirely escape attention were they quiet. A mouse could remain in the corner of a room for a long time without attracting the eyes of the occupants of the room. Let it but scamper across the corner, and at once it is discovered. It is quite conceivable that early animals were divided in the matter; that the impulse of some was to escape from danger, while others, frightened by the presence of the enemy, remained absolutely still. Each plan has succeeded. Those which, on running, ran fast enough to escape became the parents of others like themselves, led eventually to a line of animals in whose speed lay their safety. Those, however, which attempted to escape, and failed because they were not swift enough, had their line cut off, and were thus less likely to be represented in the following generation. The constant result of errors along this line would be to destroy the slow and preserve the swift, and in the course of time it is quite thinkable that only the swift should remain. As the movements grew more and more keen, even the slower of these would pass out, thus tending always to produce the succeeding generation from those who were most rapid, and hence most likely to transfer to their children a similar power.

But there is another tendency of animals which leads them when frightened by their enemies to remain quiet. If this impulse is obeyed thoroughly enough, it is easy to see how the owner of this habit might entirely escape detection by his enemy. Any restless animal unable to restrain his nervous agitation naturally betrays his presence and is picked off. The result of evolution along this line would be the exact reverse of the preceding. Those that lay most absolutely quiet would be the parents of succeeding generations, while those who were slow in coming to rest, or were indifferent about remaining quiet, were picked off, and their tendency eliminated from the future of the species. In this way many animals have come to keep entirely quiet in the presence of danger. It is not a sign of high intelligence. As a matter of fact, it is rather a stupid procedure, so far as the animal itself is concerned, but it is a preserving stupidity, and many animals have it.

The "June Bug" (which is not a bug, but a beetle, and arrives in May) has this interesting habit of keeping quiet. If in its flight it strikes the globe of an electric light, it falls at once to the ground, and remains perfectly quiet for a time. After a short interval it recovers and starts out to regain its previous activity. But this recovery is by slow stages, and the whole procedure on its part looks exceedingly stupid.

The little snake with flattened and expanded head, known as the blowing viper, or puff adder, is one of the most amusing representatives of the tendency to "play dead" that could well be found. If you strike him the faintest blow with the lightest stick, he at once goes into apparent convulsions, in which he seems to suffer the greatest agony. Then, throwing himself upon his back, he, to all appearances, yields up the ghost. If, however, you retire but a slight distance and keep your eye upon him, you find that his ghost returns after a comparatively short absence, and he slinks away out of danger. This is the most effective exhibition of this kind with which I am acquainted.

As for the habit of "playing 'possum" on the part of our opossum, the trick would seem to be particularly inane. The truth of the matter is, what is attributed to an unusual brilliancy on the part of the creature is positively unusual witlessness. The animal has an exceedingly small brain, as compared with that of a dog of similar size, and to anyone who knows brains at all this particular organ would not be looked upon as furnishing its owner much ability. The fact is that the opossum has exceedingly small wit, and this little deserts it in an emergency, as a result of which he grows helpless and motionless. This is often supposed to indicate great wisdom. There may be wisdom in it, but it is the wisdom that lies back of all nature. It certainly is not the wisdom of the opossum.

Man himself possesses to a marked degree this impulse to keep quiet in danger. The man from the country who is visiting the large city, suddenly startled by the "honk" of the auto horn, finds his power of movement promptly arrested, and he is not unlikely to be struck and injured by the machine from which the city dweller would easily escape. This is not particularly to the credit of the city dweller, who, when in the country, may find himself similarly startled by the sudden appearance of the calf, the pig, or the sheep. The bull, which a country boy, accustomed to him from childhood, will drive with a willow switch, is a source of terrified concern to the city man.

While the trick of keeping quiet serves many an animal in time of danger, there is another device for escaping attention, far more common and widespread throughout the animal world. The eye does not easily see an object if it is colored like the background against which it stands. A host of animals find their main safety in being indistinguishable in color from the surface on which they live. There are many biologists who seriously question whether protective coloration, as Darwin called it, is as effective as he believed it. In some quarters it is the present fashion to doubt protective coloration entirely. No one has yet shown any principles which will better explain the great color scheme of the animal world, and until such explanation is forthcoming I believe it will not be wise for us to discard the idea of protective coloration. No doubt it has been overworked by enthusiastic believers in its efficiency. At the same time, to overlook it completely, is, I believe, to make a greater error. I have little doubt that when the broader explanation comes, which will satisfactorily explain the color scheme of the animal world, the idea of protective coloration will be found, not so much to have been wrong, as to have been but partial. It will be included under the broader principle which takes its place and will not be supplanted by it.

The idea of protective coloration is that very many animals have ordinarily come to be colored like the background on which they live. The process has taken many generations, and is very slow, but is none the less sure in the end. In most cases the animal is probably entirely unconscious of this point in its favor, and usually it does nothing to assist the deception. The result is none the less effective because the animals themselves are unconscious of the process. The cabbage worm is green in color like the cabbage. This does not mean that it got green by eating cabbage or by longing for greennesses. Through long years the enemies of the cabbage worm have been picking it off the plants on which it fed. This does not imply that cabbages as we know them are very old, but cabbage worms doubtless ate the leaves of the sea-kale long before man had cultivated it into cabbage. During all these years the enemies of the caterpillars, generally in the shape of birds, have been assiduously gathering them up.

When we see how much the various members of the same human family may differ in complexion, how much the various pigs in the same litter may differ in size and in coloration, it is easy to understand that among these caterpillars which have eaten the cabbage there must have been considerable color variations. I do not imagine for a moment that the birds had any preference for any particular color in their cabbage worms. They took every caterpillar they saw, but they naturally first saw those that were least like the background on which they lived. The only caterpillar which was effectively hidden from his enemy was the one that was indistinguishable on the leaf. If it escaped in this way, the probabilities are that it would produce young which would be at least a little more likely to be green in color than the progeny of its darker-colored brothers and sisters. By this continued process the birds steadily weed out the darker-colored specimens. There would result, in the course of time, a race of caterpillars, whose ancestors for so many generations back had been light green in color, that there is little likelihood of any of the older and darker forms turning up again. In the course of time all dark tendencies will have disappeared from the family and practically all of the group will be light green. Any sport or variation in the shape of greater conspicuousness would fall a quick prey to the enemy and its line be cut off forever.

The same sort of activity has resulted in the peculiar green color of the katydid. This creature lives chiefly upon the leaves of trees and shrubs. This insect is so large that, even though it is leaflike in color, it might still be conspicuous. As a result those katydids whose wings were most like leaves in form were least likely to be picked up by the passing bird. This sort of protective appearance is intensified by exactly the same means as that which brought about protective coloration. The katydid least leaflike in appearance was eaten first. Thus those most leaflike remain until the last, and are most likely to produce young. Again, it was not the fact that they lived among leaves which made them look leaflike, but it is because they look like leaves that they escaped being devoured.

The katydid has materially assisted in its own preservation by being active chiefly at night. In the daytime it keeps comparatively quiet. Thus seated upon a twig, especially if hidden among the leaves, it is almost unnoticeable. At night, however, it moves about more freely, seeking its food and eventually its mate. At such times it becomes distinctly more conspicuous because its wings are steadily fluttering. The hind wings are filmy and are very light green. The creature looks most ghost-like as it flies through the evening air.

A very similar history lies back of the coloring of the ordinary toad. Though descended from the frog, and originally a creature of the water, the toad has long since adapted itself to live upon the dry ground. It still produces its young in the water as it did when a frog. Whereas the childhood of the frog, that is, its tadpole stage, is very long and it assumes its adult form comparatively late, just the reverse is the case of the toad. The young hasten through their tadpole stage within a few weeks, and assume the shape of the parent toad when about big enough to cover your little fingernail. Now they leave the water and seek dry land. Naturally they make the change when the land is damp, that is, after a warm spring rain. People seeing these multitudes of little toads hopping around over a bare spot of ground, and remembering the rain of the night before, insist that it has rained toads. Of course it never rains down anything which cannot evaporate up. The stories of showers of toads and of earth worms, with an occasional fish, or even creatures of larger size, are all pure myths. There are conceivable tornadoes after which there might be a shower of such creatures, but at such a time it is likely also to rain barn roofs and buggies. You may be sure that toads which come down in the rain are dead after they strike the ground.

The little toads started out, perhaps a hundred at a time, from the small pool in which their eggs were laid. These creatures find dragons on every side. The gartersnake comes along and gets his first toll; the heron follows him and takes such as catch his hungry eye; the turkey gobbles up his from what are left. By the time the toad-eating creatures in the neighborhood have taken such as they found, there are very few remaining. These doubtless have been left for a very good reason, generally because they were not noticed. This was because they looked like the ground on which they sat, and because they kept perfectly quiet while the enemy moved about. This process has gone on so long that the toad has come to be astonishingly well protected by its resemblance to the ground. This likeness it intensifies by its interesting habit not only of keeping entirely quiet, but of dropping its nose to the ground, instead of sitting high on its front legs, as it does when not in danger.

I have noticed that if a snake and a toad be placed in the same cage, when the snake approaches to capture the toad the toad drops into a squatting position, and is very likely to blow himself up until he is rounder in outline than he was before. Whether this is a deceptive trick which makes him the more resemble a stone is more than I can say. I do not remember having seen our eastern toad do it. I have seen it happen a number of times in the laboratory of a Colorado naturalist, and it is quite possible that in the open country more sparsely covered with vegetation than is our ground in the east this inflating device may serve the toad more effectually than if it kept its own outline.

Even among creatures far more active than the toad and the katydid an inconspicuous color must certainly result in distinctly better protection. Everyone knows the jay and the cardinal when first he has seen them, if only he has a slight acquaintance with their pictures. They are so conspicuous that we recognize them at once. More common in my region than the jay or the cardinal is the red-eyed vireo. This creature moves industriously in and out among the leaves of our trees. It is persistently in motion, is nearly constant in song, and is a bird of fair size, being larger than our English sparrow, though smaller than a robin. Many a nature lover will recognize twenty-five or thirty birds at sight without any difficulty, and not know the vireo. Yet the vireo is more common than two-thirds of the birds he knows. There can be but one reason for this; the bird is inconspicuous. The olive-green of its back, with its light under parts, serves to hide it completely amid the foliage. Even the bird-lover learns to find it first by its jerky song, and then by watching for its movements among the leaves.

One aspect of protective coloration has been brought to our attention by the artist, Mr. Abbott N. Thayer. He first clearly explained why it is that animals are usually so much lighter on the under side than they are upon the upper. Mr. Thayer proves his position by taking some ordinary cobblestones and painting one of them a uniform color and placing it upon a board painted the same color. One would think the stone would be inconspicuous; as a matter of fact, is quite easily seen. The underside of the stone, turned away from the light, is so shaded as to mark a distinct boundary between the stone and the board. Another cobblestone was colored on its upper side like the board, but the color faded into a lighter and lighter tint until the bottom of the stone was nearly white. This stone, placed upon the board, was at a short distance nearly invisible. In other words, although the pigment was actually lighter on the under side, it was so much less intensely illuminated, that the result was the same in tint as the other side under the clear sharp light of the sky.

Many a person, looking down into the water from a bridge, sees nothing whatever of the fish in the water below, because their backs are exactly like the bottom of the stream. Suddenly one of the fish, by a quick movement, turns its lighter under side over in such a way that it is clearly illuminated from the sky. Immediately a flash as of silver strikes the eye of the onlooker and makes him aware of the presence of the fish which had previously been undetected. If rendered thus suspicious, the observer will carefully examine the bottom of the water, he may quite likely find dozens of fish which had previously escaped his attention.

Nature is very versatile. So many of her apparently chance ventures have proved successful that she has retained many devices by which her children may be safe. One of these, which is doubtless often quite effective and may serve to save an animal's life, is that of being able to emit an odor so nauseating as to offend the enemy's sense of smell, and doubtless remove the keen edge of his appetite. It is not uncommon among the group of insects properly known as bugs to possess an exceedingly unpleasant odor. Anyone who has handled a squash bug will know exactly what I mean, and there are other members of the group not so common as the squash bug, which, at least to the human nose, are distinctly offensive. Some of the beetles also save themselves by this device.

One of the most interesting developments of this peculiarity is found in the case of the common skunk. This creature has in each groin a gland capable of secreting a highly offensive fluid. Ordinarily this liquid is kept safely within its sac, and for a long time none of it may escape. When closely cornered, the skunk will turn its tail toward the enemy and with a quiver and a flip of his tail it can guide the openings of two little tubes that come out along the root of the tail in such fashion as to eject the fluid in a fine and foul-smelling stream against the animal from which the skunk would escape. Once fairly hit by this fluid, I imagine most animals will drop the skunk. A dog surely will, and will hate himself for having made the attempt to capture anything which must be so ignominiously allowed to escape. If ones clothing is well saturated with it, it is nearly useless to hope to remove the odor. A dog will carry the smell for several weeks. For a long time it will be so strong as to make him an unfit denizen of the house. Even swimming in deep water does not remove it. After two weeks, although he may seem to be practically free from the odor, a light rain will bring it all out again and make him nearly as offensive as before.

Not as prompt in its action, but in the end nearly as effective, is the protective device which the toad sometimes uses to his distinct advantage. May I be pardoned a personal account of this particular feature. It was my good fortune to be for a short time a student in a class taught by Edward Drinker Cope, one of the most brilliant of our American biologists. Prof. Cope mentioned in class the fact that the Batrachians (the group to which the toad belongs) have in many cases the power to emit from their skin a fluid which is sufficiently nauseous to deter an animal from eating the creature that secretes it. Upon such authority as this, I had no hesitancy whatever in repeating Cope's statement. One morning I had a class in the field studying the ground ivy, whose dainty blue flowers were lifting themselves out of the dewy grass. While we were thus engaged, a toad joined the circle. He came out of his dewy retreat clean and fresh from his morning bath. I took him in my hands, and made him the subject of an immediate lesson. I showed to my pupils his eyes and his interesting method of handling them, his tongue and its strange insertion; showed them how to look into his mouth and look up his ears to his ear drums, and pointed out many other interesting facts. Then I told them how Cope had said that the toad had power to emit from its skin a fluid so nauseous that many an animal hesitates to eat it. This is the first peculiarity I had mentioned which I had not myself observed, and a scientific qualm came over my conscience. Why had I never verified this statement which I had so frequently repeated? On the impulse of the moment, with the bright, clean skin of the creature fresh from the dewy grass, making it less than usually repulsive, I ran my tongue up its back only to find that it had no taste whatever. I was of course surprised, but I was not foolish enough to deny, as the result of one observation, the statement of a good scientist. The observation, moreover, was one which I naturally did not care to repeat with any frequency. Of one thing I was sure, toads do not always have an unpleasant taste.

A year later I had a class down by the side of a neighboring pond. The pool was not an attractive one, and I had picked from it a more than commonly unappetizing looking toad, which proved to be a mother which had not yet laid her eggs. As I held her in my hands and exhibited her various points to my pupils, I told them of Prof. Cope's statement. I also told them of my unsuccessful attempt the previous year to verify the statement. I added, however, that I would not repeat this experiment on this unappetizing specimen. Hereupon the toad not only exuded, but squirted, from a gland over her left shoulder blade a fluid, milky-like in appearance, and forming a jet as thin as a needle, but ejected with force enough to strike my face, which was at least fifteen inches away. I moistened my finger on my tongue, lifted the fluid from my cheek, and tasted it. Cope was right. A toad can exude a most nauseous fluid. Horsechestnuts extracted and distilled might possibly provide something as bitter. Why did I not find this in the preceding case? I have too few observations on which to base a conclusion, but I have a suspicion as to the reason. In the case of the toad which spurted the fluid in my face, we had a creature with whose life were tied up the lives of her many offspring, to be produced from the eggs she was so soon to lay. Under conditions like these, nature is more than commonly careful of her children. Whether this be the reason or not, toads do not always have an unpleasant taste, but when they do it certainly is most unpleasant.

There remains to be considered the most effective plan yet mentioned of escaping the enemy, and that is of really escaping. In all the devices we have considered thus far the enemy is eluded. When the creature lies quiet, or finds safety in its protective coloration, or in its bad taste, or unpleasant odor, it still remains in the presence of the enemy. A more progressive plan altogether is to escape the enemy by flight. The great advantage of this plan lies in the fact that the acquisition is valuable for every purpose. The creature then can escape the enemy, can range widely for food or for a mate. This gives it an enormous advantage in the struggle for life. The power to fly, in insects, was doubtless originally gained in the attempt to escape the enemy. Among many of the lower animals it is nearly the only purpose that flying serves. Later on it enables the animal to pass from one food locality to another. In a few creatures it plays an effective part during the mating season. These last are probably both derived powers, and the original function was that of escape from the enemy. The grasshopper has grown its long legs to serve him for safety, and through them it is helped along, moving about chiefly by leaps when it wishes to go any material distance. It is only toward the very end of its life that the grasshopper has wings, and then they serve probably to aid in the search for a mate. Among the birds flight began simply in sailing out of the trees, into which the creature, still half lizard, had crept to escape its enemy. The earliest bird known to us had comparatively insignificant wings. There was really more support in its tail than in its wings, and this would distinctly indicate that it glided more than it flew. It had claws also upon its wings, and it was probably the case that this creature crept into the trees, at least in its earliest forms, and sailed down in a manner not unlike that employed to-day by the flying squirrel. From such simple beginnings came the wonderful power of flight in the birds.

Among mammals the attempt to escape from the enemy has led to an interesting development, which will be more fully explained in a later section when we speak of the history of the horse. The early mammals walked flat-footed, as we do on our feet and as the raccoon and the bear do on theirs.

Gradually, however, as their enemies became more fierce and better able to injure the larger mammals, the latter gained in power of flight, and this gain consisted first in rising from the toes, lifting the heels completely off the ground. At the same time the leg and foot were gradually lengthened. Doubtless in this way the fleet animals, like the deer, the horse and the giraffe, first came by their long legs. Constant elimination of the short-legged ones, by the pursuing enemy, resulted in the selection of the long-limbed ones for breeding purposes, and hence to the ultimate elongation of the legs of the species.

The method of escape from the enemy involves cowardice. "He who fights and runs away may live to fight another day," and so it may be the part of wisdom in the weak creature to escape from his enemy by flight. It is a far more estimable process, from our standpoint at least, to stand against the onslaught of the enemy and beat him upon his own ground. This end is secured in many animals by acquiring horns or by lengthening certain of the teeth. The horn is a very ancient instrument of defense. When the reptiles ruled the land horns were not uncommon. They consisted in those days of hardened scales, which lengthened and fastened themselves over a core of bone. Such an old-fashioned instrument, sometimes made of newer materials, still remains the defense of a number of animals. The rhinoceros has upon his nose a lengthened projection, which is what might not improperly be called hair glued into a cone. This enormous horn is a frightful weapon, both of offense and defense, and, when backed by the terrible weight of the body of the rhinoceros, it can do as deadly work as almost any instrument of destruction known to animals below the grade of man. But, after all, this is an old-fashioned method, and the rhinoceros is a relic.

Among the carnivorous animals the teeth, which were developed first chiefly for the tearing of flesh in its consumption, became effective for their courageous owners. Because these tearing teeth are well developed in the dog they have come to be known as canine teeth. Usually where an animal can use its teeth effectively for offense or defense, it is the canine teeth that are thus modified. The cat has developed them better than the dog, and one of the cats of a bygone geological period had canine teeth so magnificently enlarged and so sharp at the back as to give this frightful creature the name of the saber-toothed tiger. The long teeth in the upper jaws of the elephant, commonly known as tusks, are not canine teeth. The elephant has completely lost his canines. His tusks are his incisors, and they have developed as have almost no other teeth in the mammals.

These are only a few of the numberless devices nature has evolved for furthering the success of her children. There are so many others that to many of us they form almost the chief point of interest in our study of a new animal, or our closer observation of an old friend.

CHAPTER V

ADAPTATION FOR THE SPECIES

The strife, as we have described it thus far, is a purely selfish struggle. Every point gained is a point favorable to the welfare of the individual animal. But nature is uncommonly careless of the individual unless the advantage gained is also of use to the species as a whole. Very often the life of an animal ceases when provision has been made for its young. The male garden spider may have a long and dangerous courtship, in which the uncertain temper of his ladylove may lead her to bite off four or five of his eight legs. But her ingratitude is not yet complete. He may have barely accomplished his desperate purpose of fertilizing her eggs at all hazards, when she ends the process by eating him. The male bumblebee fertilizes the female in the late summer and then dies. She does not lay her eggs before the next season. So it happens that no bumblebee ever sees its own father, and no father bumblebee ever sees his own children. In the honey bee the male, which has been fortunate enough to fertilize the queen, pays for his honor by death within the hour. Superfluous bachelors, among the honey bees, when the bridal season has passed, are driven from the hive to die of starvation.

An animal need not always be successful himself, but it is more essential that he hand down his successful traits to those who come after him. It is more important for the future generation that an animal should have had it in him to do great things, though he himself really have never done them, than that he should have learned to do great things on a meager original endowment. Not what an animal accomplishes is important to his children, but what he has it in him to accomplish. Accordingly Nature is full of devices by which those who have proved their original endowment by winning out in the struggle shall hand on this endowment to a subsequent generation. In other words, Nature is anxious that they may successfully mate. Here we are again on distinctly debatable ground. Darwin himself believed thoroughly in what he called sexual selection. It is the essence of this idea that the males and females have grown unlike, more technically have developed secondary sexual characters, through the choice of the mating pair. It would usually be the more serious loss if accident should come to the female, for she may carry fertilized eggs for some time. Hence, if both sexes may not become attractive, it is usually the male that develops fine colors, ornamental appendages or a captivating voice.

An interesting reversal of this process has taken place in civilized man. His more savage ancestor adorned himself more lavishly than he permitted his mate to do. With the advance of civilization man has undertaken to defend his own mate most valorously. The result is it is safe for her to be beautiful. Under these circumstances, however, it is more necessary to her welfare that her consort be vigorous rather than that he be handsome. Hence in the human species beauty has become the prerogative of the woman, and this is increasingly the case the higher the civilization. Whether woman suffrage and self-support will reverse this process remains to be seen. There are indications that point that way.

There are many biologists who are at present expressing serious doubt as to the validity of sexual selection. As in the previous cases of protective coloration, I believe it will be wise for us to retain, even though with an interrogation point behind it, the idea of sexual selection until such time as those who object to it have furnished us with another theory which will more nearly account for the observed facts. While entirely conscious of the possibility that there is a weak spot in the theory, we will still tentatively hold to sexual selection. The fact that beauty in women is so intensely attractive to man, and that vigor and manliness in man are so attractive to women, leads us to infer that among the lower animals, although of course in a vastly less degree, vigor and beauty are also attractive. The weakest point of the position lies in the fact that it probably presupposes a higher degree of capacity for appreciation on the part of lower animals than they possess. Those who deny the truth of the theory laugh at the idea that a butterfly can see clearly enough and care enough for what it sees to notice whether its mate has wings of one type or of another. The size, number and position of the spots on the wings of many butterflies are so nearly constant that they cannot of themselves have been entirely determined by the choice of the insect. Yet this may not preclude the possibility of the fact that, while the spots were produced through some other agency, certain types of them were selected by sexual preference.

If attractive coloration is effective anywhere in the animal world, it will possibly be found among the insects, but it is especially likely to be found among the birds. Very many field workers in these groups feel quite sure of the value of attractiveness. When butterflies chase each other up and down, circling and doubling, following each other for long distances, it would certainly seem as if they were pleased with each other's appearance. Some naturalists, especially those who have worked chiefly in the laboratory, insist that it is the odor, not the color of these insects, which is attractive, and some experiments which have been made would seem to point in this direction. But the creatures experimented upon most carefully were night-flying moths, and it is quite possible that the sense of sight in the night-flying moths has lost its vigor.

The great difficulty in understanding sexual attraction in insects, as based upon beauty, lies in the undoubtedly lower development of their nervous activity; in other words, in the apparent absence of anything worth calling mind. I think no one imagines that a butterfly, looking upon two other butterflies who are competing for her affections, deliberates between them and determines to admit to the circle of her friendship the more brilliantly colored male. Moths are so irresistibly attracted to a light as to fly into it without apparent power to withstand its influence. They repeat the flight again and again until they are destroyed. If they react so vigorously to the stimulus of the light, it seems not impossible that they may also act vigorously to the stimulus of color pattern, and that the male most beautifully colored, according to the nervous ideal of the female, should win her unconscious regard. At least it is certain that, in very many of the butterflies and moths, the attractive coloration is chiefly displayed when they are moving actively about; and when they alight and their enemies can the more easily capture them, they conceal their brilliant colorings. Most butterflies are very brilliant on the upper surface of the wings and very much duller on the under surface. Hence in flight they show their colorings exquisitely, but when they alight, and are thus more likely to be captured, they fold the brilliant surfaces together in an upright position. In this way not only is the dull side of the wings placed outward, but the wings themselves are placed edgewise to the sky, and it is from this direction that their enemies, the birds, are most likely to see them. Once upon the wing these creatures display their beauty with much greater safety because they can escape the birds very readily by use of their exceedingly jerky flight. The butterfly's motion is as irregular as any we have except the bat's. This eccentricity is one great element in their safety, and makes it less dangerous for them to display their attractive colorations.

One very large group of the night-flying moths have been named the "underwings," because of the fact that their hind wings are very much more brilliant than the front, and in lighting they fold the dull pair back over the bright, completely concealing them. These creatures are in the habit of resting in the daytime against walls, or stones, or the bark of trees. The similarity in color between their front wings, which alone show while sitting, and the background on which they rest, is most remarkable. One may pass them again and again, although they are of considerable size, and not notice them at all. Once let them display their hind wings and the brilliancy of their color always attracts immediate attention.

It is among birds, however, that brilliant coloration serves its most effective purpose. The birds are alert, exceedingly quick of sight, and are much more discriminating than insects in almost every respect. It is not so impossible that these creatures might even voluntarily prefer a distinctly more brilliant mate, though the voluntary character of the process is not essential to its success. Men certainly are constantly attracted to women for whose charm it would puzzle them to account. If this is true with regard to men, it is certainly probable that birds would be largely influenced by phases of attractiveness, of which they were observant, but unconscious.

Certain it is that in many birds the males are far more beautiful than the females. Perhaps the commonest illustration, and, at the same time, one of the best is found in the so-called red-wing or swamp blackbird. The male of this creature is a brilliant black, excepting that upon the angle of the wing, spoken of roughly as his shoulder, though in reality it is equivalent to our wrist, there appears a splendid orange patch with a border of lemon yellow. When he folds his wing he pushes this colored angle of the wing so deftly under the feathers of his shoulder as almost to conceal it. When in flight the bird is exceedingly conspicuous, showing, with every bend and twist of his body, his gorgeous epaulets. Meanwhile, the female is likely to pass unnoticed. She is dull in color and streaked like the grass among which she lives. During the mating season the male hovers about her, swaying from side to side in such a way as certainly to make it appear as if he realized his good points and was bringing them to bear as effectively as he knew how. After his mate has nested and is rearing her young, it would appear that the male uses his brilliancy to lure the observing enemy away from the nest containing his wife and children.

Another illustration of the remarkable superiority of the male over the female, in many parts of the bird world, is seen in the case of the common barnyard fowl. The rooster is so much more gorgeous than the hen that anyone reasonably acquainted with these birds cannot have failed to notice the fact. In some of our modern varieties we have by breeding colored them nearly alike. The original chicken is colored much like the common Leghorns. Shades of red and yellow decorate his neck and back, while the flight feathers of his wings and of his tail and the sickle feathers which ornament the rear of his back and hang over his tail are lustrous dark green. The hen meanwhile is very much less brilliant in her contrasts. I shall speak more fully of this in discussing polygamy.

The attraction of beauty is not the only lure by which a creature may win its mate. Sound may captivate as effectively as beauty. This is true of insects as well as of birds. Certain insects at least advise their mates of their presence by means of a sound which they emit. This is particularly noticeable among the group of straight-winged insects to which the grasshopper, katydid and cricket belong. The grasshopper has a ridge on the angle of his wing and a roughness on the side of his leg. When these two are rubbed together the result is sometimes a fiddling, sometimes a snapping or cracking sound, differing in different grasshoppers. I doubt not these sounds are pleasing to the female of the species, for they are always made by the male. The katydid, instead of fiddling in this way, has a sort of drum on the angle of his one wing, which he can rub over a tooth in the corresponding angle of his other wing, thus producing the familiar "katydid" sound. I have never succeeded in making a dead grasshopper fiddle, but I have long known how to make a dead katydid say "ka." Quite recently I have added to my accomplishment in this respect and can make it say "katy." The "did" part of the song still lies beyond my power. The crickets produce their sharp notes in much the same fashion as the katydids.

One observer of the chirping of the cricket says that the pitch of the song varies with the temperature. He has even worked out a formula by which one can tell the pitch of the chirp, if he knows the temperature, or, knowing the temperature, can determine the pitch. Of course this is too mechanical; yet it indicates that there must be considerable relation between the two; the warmer the cricket the happier he is.

It is the males among insects that chirp their love songs. The females never answer them. There is a peculiar notion that the female katydid, when thus accused of some offense, replies "katy didn't." The truth of the matter is that no female katydid ever replied to the accusations of her lover, if accusation it be. She is absolutely dumb, not having the drum upon her wings with which to reply. She is provided with ears wherewith to hear, and, strange to say, she keeps them on her elbow, as does also the cricket, while the grasshopper has his ears upon the side of his body.

Everyone who lives in the country, or goes into the country in the summertime, is sure to know the humming of the so-called locust. It is an unfortunate fact that the word locust may have several meanings. It is properly applied to one group of the grasshoppers. The creature most commonly called a locust is a cicada, or harvest fly. When the weather gets quite warm the cicada starts his love song. He has two long flaps to his vest, and under each flap he has a vibrating drum head. This is set shivering by a muscle on its under side. The female cicada again is silent.

It is among birds that the love song reaches its finest development. It may consist simply of a little chirp as in the chippy. It may consist of two notes of a different pitch repeated steadily, as in the tufted titmouse. It may attain considerable variation, as in the robin. But in the choir of our best singers, like the catbird, thrasher, and mocking bird, there is unending variation of notes. It seems almost impossible to doubt the charming quality of this voice upon the mate. It certainly is chiefly confined to the mating season, and is indulged in almost entirely by the males. This does not mean that a male does not sing excepting when he wishes to charm his mate. But the time when he is in his most exquisite feather and most charming mood is the time when he sings most sweetly, and this is the time when he is taking to himself a mate. The love joy may so overcrowd his life that he sings much and often, but the increase in its amount and character during the mating season seems to proclaim its purpose beyond a doubt.

In addition to the allurements above described there are certain peculiar behaviors of the animal during the mating season which are intensely interesting. Sometimes they consist simply of a wild delirium of joy, which overpowers the animal completely and makes him do wonderful things. Birds will fly with impetuous leaps in the air, mount higher and higher, singing wildly, only to turn suddenly at the top of the flight and drop promptly to the ground. I have seen such ecstatic flights in the oven bird and in our rollicking gold finch. I have seen a catbird on his way to a tree turn three somersaults, much like those performed by a tumbler pigeon, after which he alighted upon the bough. None of these acts seemed deliberately performed in front of the females, but I have seen three or four killdeer parading in most stately and precise manner, spreading their wings and fluffing their feathers, performing a sublimated cup-and-cake walk amid a circle of attracted females.

Even our little English sparrow, as I have previously mentioned, fluffs himself up and spreads his wings and prances around in front of his presumably adoring ladylove. But the weirdest performance of this sort I have ever seen is that shown by the male ostrich. When he becomes excited, swaying his body from side to side, he sinks slowly upon his knees, until his body touches the ground, his wings spread on either side and the feathers fluffed up so as to show every exquisite plume in all its splendid beauty. The long neck is laid back until the head, which is doubled sharply forward, is pressed almost against the back, and in this strange position he sways from side to side, apparently utterly oblivious, for a time, of everything. After about a minute of this performance, he seems slowly to come to himself and rise again to his feet. Now he is particularly likely to make vicious attack upon anything within reach.

It is not only necessary that the animal should be able to attract a mate. There may be more than one claimant for the damsel's affection. In many animals we see provisions whereby the male may effectively deal with his rivals. This is especially likely to be the case if the animal be a polygamist. In every species there are produced about as many males as females. If the polygamous habit leads one male to gather about him a group of females, with whom he mates, it is evident that he is displacing an equal number of rivals, and they are not willingly displaced. Accordingly we find that polygamy is usually accompanied by a belligerent disposition on the part of the males. In our ordinary barnyard fowl this trait is very evident. The rooster not only domineers over the hens, not only struts about among them in stately fashion and gives vent to his feelings by his sonorous voice, he must also drive away from the neighborhood any rivals for the affections of his wives. Hence the rooster attacks upon sight the neighboring rooster, and battles with him to his entire discomfiture and sometimes to the death.

Among the members of the deer family this particular phase of the relation between the sexes has produced in the males, and only very rarely in the females, the magnificent branching horns. These are intended not so much as a protection against the enemy as for an offensive weapon in the battle for the mates.

Beautiful and stately as are these magnificent horns, they last only for a part of the year. We begin to understand their meaning. When the wolf is hungriest, toward the close of the bitter winter, the deer is without horns. When the time for mating comes, the deer within a few weeks grows his horns, which at first are covered with a plushlike coating, known as velvet. After a while this dries and he rubs his horns against the trees until they are clean and smooth. Now he is ready for the battle royal.

In the case of the fur seals polygamy has carried its specialization of the males to a remarkable extent. The bull seals are several times as large as the cows, and are provided with terrific canine teeth. With these they battle with a violence that very often results in the death of one of the combatants. A successful bull seal who has gathered about him a cluster of seal cows is seamed and scarred with the marks of his annual combats.

One more type of adaptation can be profitably considered. Animals have developed many devices which serve for the protection of their young. The wonderful silk spun by the spider was evidently primarily intended to serve as a covering for the eggs. Probably all of our spiders agree in using the silk for this purpose. Many of them employ it for practically no other, though there are half a dozen different uses to which different spiders may put their silk.

Under these conditions we have a right to infer that silk was primarily developed as a coating for the eggs. In the case of some of our spiders a little fluffy mass of silk covers the egg, while a firmly woven sheet of silk covers both egg mass and fluff, holding it flat against a wall or the trunk of a tree. In some of the higher spiders, notably our bank spiders, the silken covering becomes an effective cocoon, spherical in shape, with a little opening at the top like the neck of a small bottle. The egg cocoon is woven in a mass of tangled silk between the branches of some tough weed which will be sure to outlast the winter. Into the egg cocoon the spider may place one thousand or more eggs. Having thus provided her children with a snug winter home, the spider dies. When spring comes with the warm rays of the sun, the eggs hatch and the cocoon becomes a creeping mass of minute spiders. At the time these spiders appear there is nothing for them to eat. The obvious way out of this difficulty is taken. At once there begins a progressive party. Spider fights with spider, and the prize in each conflict is the body of the victim, which is promptly eaten. The winners in the first round pair off again, and a little later, as hunger drives them, another set of combats comes on, resulting in another halving of the number of spiders in the cocoon. This process continues until not more than one-tenth of the original number of spiders remains. By this time they have gained sufficient strength of leg and jaw, and sufficient dexterity in the use of both, to make it safe for them to venture out and try their fortunes among the accidents of a strenuous world. There can be little doubt after this long process has worked its final results which tenth remains. Chance plays but small part in this game. It is the fittest that survive. When this procedure goes on generation after generation, the result must necessarily be that the spiders grow fitter and fitter for their work. This method is hard on the little spider, but it makes good spiders.

Most insects die before their eggs hatch; accordingly they can pay no attention to their own children. Whatever arrangements are provided for the safety and strength of these offspring must be provided before they appear. About the only care the majority of insects take in this direction is to see that the eggs are placed where the young shall find food as soon as they emerge. Insects' eggs are very small, and as a consequence the creatures which emerge from them are likewise exceedingly minute. As a result they cannot be expected to hunt far for their food. Different insects use different devices by which to overcome this difficulty. The katydid, for instance, must die with the approach of fall. Her children will not appear until the following year. Her food consists of leaves, but to lay the eggs in such a situation would be a fatal process, because the leaf will drop off before the eggs hatch. Accordingly, the katydid lays its shield-shaped eggs in a double row near the end of a young twig. Next year when the weather is sufficiently warm to hatch katydids, it is also warm enough to force the buds on the end of the twigs. When the katydids arrive their jaws are young and tender, but so are the leaves upon which they are born. Hence there is little difficulty on the part of the young katydids in finding an abundance of food. By the time the leaves have grown tougher, the katydid's jaws are stronger, and the leaves will still serve as food.

Everyone who is at all familiar with country life and gardening is familiar with what is called the potato or tomato worm. It is a long, green, smooth, caterpillar, as long and as fat as your finger and provided with a horn upon his tail. The gardener may not know that after a while this creature will burrow into the ground, and there change into an oblong brown mass with a sort of a pitcher handle at one side. Next year this pupa will split down the back, and from out of the brown case will come a hawk-moth, which soon will fly with rapidly quivering wings and feast upon the nectar of our moon flowers or on that of the "Jimson" weed. Those who have cleaned these pests from the potato or tomato vines will often have noticed one of them covered with what look almost like grains of rice. This appearance reveals an interesting story. Some time earlier an insect that looked very much like a dainty wasp with a rather long sting in its tail hovered over the caterpillar. This is the ichneumon fly. Eventually lighting upon the caterpillar's back, it punctured the skin with its sting, and deposited eggs within the caterpillar's body. These eggs soon hatched and the little grubs worked their way through the body of its host. The infested victim feeds upon leaves and fills itself with rich food. These parasites eat the food, and, try as it may, the caterpillar does not succeed in getting fat. After the grubs have gotten their full growth, each of them eats its way through a little hole to the outside of the caterpillar's body. Here it spins around itself a little white case, and looks like a rice grain. As the caterpillar moves about, these seeming rice grains are rubbed off and fall to the ground. Next year there will come up new ichneumon flies to sting fresh caterpillars and repeat the entire process.

Another remarkable provision for the young on the part of insects is seen in the behavior of the big sphex wasp, known as the cicada killer. The cicada, it will be remembered, is what is commonly called a locust. The cicada killer is a magnificent big wasp, whose body is nearly an inch long, banded with black and yellow, while the wings are colored a smoky brown. This muscular wasp digs a long tunnel eight or ten inches deep, which ends in a slightly larger room. Having provided the location, he now sallies forth in search of the cicada. The heavy song of the male probably serves as a guide to the wasp in case of scarcity of cicadas, but the killer has apparently little difficulty in finding his prey. The wasp pounces upon the insect, and in spite of its strength and the thrashing of its vigorous wings punctures it with his sting again and again. The poison of the sting entering into the nerve centers gradually paralyzes, but usually does not kill, the cicada. Now the killer carries its prey home, pushes it to the bottom of the tunnel and deposits upon it a single egg. The wasp closes up the hole and leaves the place. When the egg hatches and the grub of the wasp emerges, it finds a big cicada just at hand, upon which it feeds. By the time the cicada is completely devoured, the wasp grub has obtained its full growth. After a short period of development a new sphex wasp is ready to work its way out of the tunnel, find a mate, dig a hole, and safely provide for its own children.

Still more remarkable adaptations for the care of the young appear among the birds. Here the eggs are not to be deserted, but are to be cared for until the young appear. These again must have attention until such time as they are quite able to take care of themselves. The birds are warm-blooded animals, and even their young, while they are developing in the egg, are warm-blooded. Consequently the temperature of the egg must be maintained evenly and uniformly, or there will be no development.

The fish may drop its eggs carelessly upon the bottom of the stream. A frog may deposit them in a mass of jelly and leave them forever. A turtle may bury its eggs in a sand bank and abandon them to their fate. The warm blood of the young bird demands more attention than this. Accordingly, the parent bird has learned to make for itself some sort of nest, in which the young may be kept properly warm until they are developed. The ancestral bird, who was to be the progenitor of the entire bird class, must have had some very simple method of providing a place in which its eggs might be hatched. As the descendants of this original bird have passed into new situations, the various lines have taken upon themselves different shapes until we have the multiform birds of to-day. The habits of the birds have also varied. Each has adapted itself to the situation in which it found itself, and no adaptation has been more varied and effective than the adjustment of the nesting site. Nests are found upon the ground, in the bushes, on the lower limbs, in the crotches of the trees, in the trunks of the trees, upon their very summits, and on the tops of inaccessible crags. To every sort of situation some bird has been enabled to adapt itself. This has made it possible for very many more birds to thrive than could have found a place in the world, had they all lived upon the same plan.

In the case of the bank swallow his nest may be a very simple contrivance, consisting only of a tunnel running back into a bank, and widening at the back. Some material that will soften the bed upon which eggs are to be laid must be placed in this cavity. The whole home is a very simple and crude affair. But little better is the arrangement which the woodpecker calls a home. This has been cut into the dry wood of a defective tree. No woodpecker can make his home in absolutely solid sapwood. Hence the first labor of the woodpecker must consist in finding a place in which it can dig. If there is an old stump of a limb sticking up, the problem is readily solved. Such wood has no sap in it, and is brittle enough to be easily dug out. But, if there be no such stub, the woodpecker will find a suitable place in most trees. At some time or other almost every tree loses a big limb. When such accident occurs there will always be in the old trunk a region through which sap once went to this limb. This region, deprived of its function, goes completely dry, like the heartwood of the tree, and it is into such material as this that the woodpecker succeeds in drilling his well-protected home.

As birds rise higher in the scale the nest-building becomes a more complicated affair, and after a while we find a well-woven substantial nest, through which even the air will not chill the eggs enough to prevent their hatching, while the warmth is supplied by the mother's body. It is often a matter of surprise to many people that a bird should contrive to build a nest so exquisitely circular. The trick, after all, is not quite so difficult as it looks. The robin gathers up a few sticks and places them as the beginning of the platform. More and more are brought and woven into each other, making a framework altogether too big for the nest. Then mud is brought and plastered inside of this. With the plastering of this mud the careful circularity of the work begins. Every time a little material has been added the robin sits down in the nest and revolves her body, in this way shaping the interior much as the potter shapes a pot. In the case of the artisan, it is the pot that revolves. In the case of the robin, the bird itself revolves. The effect is the same in both cases--a circular vessel is produced. A little lining added to the interior of the nest softens it for the reception of the eggs. In this exquisite home the robin lays her eggs, and sits upon them until they are developed enough to hatch, and then feeds the young until they are old enough to feed themselves.

Far more remarkable than any of the devices thus far described are the wonderful developments which have come in the class of animals known as the mammals. Here the most wonderful protection is made for the care and feeding of the young. But this is to be the subject of a separate chapter.

As long as we thought of each sort of animal as being a separate species shaped in the beginning by the hands of the Creator, each of these devices seemed to us a new manifestation of the Divine Providence, whose fertile planning had conceived so many methods of providing for his children. Unconsciously we thought of God acting as man acted. Each animal seemed a purely separate invention purposely designed for an especial place. Now we understand the plan in creation better, and see that each animal has come from another not quite like itself, some distance back, and this from still another. Our admiration for these devices as they arise through evolution is no less, but takes on another form.

CHAPTER VI

LIFE IN THE PAST

Anyone who earnestly studies plants and animals as they exist in the world to-day cannot help wondering how the earth began and where it got its life. This is the true end and aim of geological study. The history of man seems to run back into a far distant and gloomy past. Except for the poetical account in Genesis and the traditions of various peoples throughout the world, real history fades away into an earlier time of which there are no written records. When the delvers in the Mesopotamian plain talk to us of kingdoms running back through seven or eight or nine thousand years, we seem to be getting back to the beginnings of things. But seven or eight or nine thousand years are as nothing in comparison with the age of the earth, which runs back into a past so limitless that no man can safely assign any set figure to it. In a recent paper, Dr. Walcott, of the Smithsonian Institution, says that the antiquity of the earth must be measured not in millions, for they are too short, nor hundreds of millions, for this carries us too far, but must surely be measured in tens of millions of years.

When we attempt to study the past we find its various epochs unequally clear to us. In human history only quite modern times are absolutely clear. The history of the Middle Ages is distinct enough for us to build for ourselves a picture of the time with reasonable hope of gaining a correct view of the state of affairs. Back of this comes the long stretch of the Dark Ages, in which here and there we have bright spots, but it will perhaps long be impossible to portray clearly the life of the people. Getting back to the Romans, things once more become reasonably plain, as is true also in the case of Greek history. Back of this stretches the Egyptian with fair precision, and, older than it, the Babylonian and Chaldean. But these past three have not left nearly so definite an account for us as did the later civilizations of Greece and Rome.

When we try to go back of these we must change our method of study entirely. Writing is absent, and all we know of earlier men must be inferred from a few pictures that were daubed on the rocks or carved in ivory or bone, from tools made of stone or bone, from a few metal or stone ornaments, or from the bones of the men themselves. Even so, the history fades out without telling us its own beginnings. It is quite as impossible for history to write its origins as it is for man, from his own knowledge, to describe his birth.

What is true of the human story is quite as true of that of the earth. Recent steps are very plain. We may read them with considerable confidence. As we go deeper into the rocks and find older fossils, the evidence becomes less certain. The animals differed enough from those of to-day for us to be less sure what they were like. As we keep on moving backward through time, and downward through the rocks, we find, after a while, strata in which there are evidences of life that existed long ago, but in which these traces are so altered that it is impossible to tell what sort of living things existed; we learn only that they were alive. Going back still further, these fade out. There is no knowing when the earth began; there is no knowing when life began upon the earth. It is not meant that men have not wondered, even reckoned carefully, as to how long ago each of these events occurred. Many speculations have proved entirely useless, a few remain as yet neither confirmed nor disproved, and of such we shall speak.

For the last hundred years the theory of the earth's origin suggested by the Marquis Pierre Simon De La Place, of France, near the end of the eighteenth century, has held almost undisputed sway among men who were willing to consider the question as open to human solution. This theory is known as La Place's Nebular Hypothesis. When men began to study the heavenly bodies with the newly invented telescope, new ideas naturally sprang up. Among the objects which the glass disclosed were the nebul? which are great clouds of fire mist, glowing masses of gas. They are scarcely visible to the naked eye, but are among the most interesting objects in the heavens when seen through a telescope. The other suggestive heavenly body was our sister planet, Saturn. Besides having a full complement of moons, Saturn has around it, as distant as we would expect moons to be, three great rings. These look very much as if one's hat, with an enormously wide brim, should have the connection between the rim and the hat broken out completely, but the rim should still float around the hat without touching it and should steadily revolve as it stood there. The rings of Saturn are not solid like the suggested hat rim. They are evidently made up of a great number of very small particles, each moving around the center of Saturn. But the great cloud of them is spread out flat. At the distance which Saturn is from the earth they look as if they made a solid sheet. Furthermore, they do not form, as it were, one continuous hat rim, but it is as if the rim were broken into three circular sections, each bigger than the one inside it and separated from the next by an area nearly as wide as the ring itself.

With such material in the heavens to guide him, La Place suggested that the sun had once been an enormous fire mist scattered over an area billions of miles in diameter. This gaseous material, by the attraction of its particles for each other, began to condense and contract. When the plug is pulled from a washbasin the particles of water, in moving toward the center, in order to get out of the basin, invariably set up a rotary motion. As the particles of this diffused nebula began to gather together they, too, gave to the mass a rotary movement. This grew more and more rapid, with greater contraction, until the particles on the outer edge of the rotating mass had just so much speed that the least bit more would make them tend to fly off as mud would fly from a revolving wheel. When this point was reached there was a balance of forces which made the outermost portion remain as a ring while the rest contracted away from it, leaving it behind.

It was La Place's idea that this process had repeated itself, and ring after ring had been left behind. Finally the sun condensed and grew into a ball, occupying the center of the system. At varying distances from it were to be found either rings or planets which had been formed out of such rings. For La Place suggested that in a ring like this the material could not be quite evenly distributed. While every particle in the ring kept revolving around the sun, those in front of the densest part were slowly held back by the attraction of the thicker portion, while those behind it in rotation had their speed hastened until finally all the material in the ring had collected at one spot and a new planet was born. La Place believed that these planets formed their moons in exactly the same way, and that Saturn was simply a planet not all of whose moons had yet been formed. He believed that this happy accident served to tell us how the universe had been created.

Of course, so detailed a theory concerning anything of which we know so little has always had much ridicule thrown upon it, and yet no truly competing theory has been proposed until very recent times.

Within a few years a Planetesimal Theory has been announced, and is gaining considerable prominence, although it is too early yet to say whether it will supersede La Place's idea. In this theory, also, the suggestion comes from the heavenly bodies. With the increasing study of the nebul? many forms of these interesting bodies have been discovered. A very common type consists of a great coherent central mass, with two or more arms extending from opposite sides in the form of a spiral. This is as if gaseous revolving nebul?had come into comparatively close proximity to a passing body. The visitor, by its attraction, drew from the nebula a wisp of gas. The revolving motion of the nebula gave to the attracted arm the spiral form.

These twisted arms are not equally dense throughout, but have thickened knots here and there in their course. The Planetesimal Theory suggests that these thickened knots are embryo planets and the central portion of the nebul?an embryo sun. After all the material in such a body has condensed either around the knots or about the central mass a new solar system will be complete. As before stated, neither of these theories can be said to be demonstrated. Each of them has points in its favor and each has its difficulties.

It is pleasant to know what men have clearly thought concerning such questions, but for a man not a trained geologist neither will carry much conviction. He will still rest with his own early conclusion that whichever shall prove to be true, for him his old formula is still valid, "in the beginning God made the heavens and the earth." He will no longer think of God as having shaped the balls with his own hand and thrown them into space; he will no longer dream that it all occurred within a week not more than six thousand years ago; but still to him will come the reverent conviction that, whatever the plan by which it was accomplished, it was still God's plan and God carried it out.

Now that we have tried to stretch our imagination back to the origin of our globe, the question not unnaturally comes to our mind, how long ago did all this happen? Is there any possible means of telling when the history of the earth began? All such attempts lead either to indefinite or to uncertain conclusions. Each man who essays the problem approaches it from a different side and ends with a different result. But no matter what the method of approach, all are agreed on at least one point, the enormous length of time, as counted in years, through which the earth has lasted.

One great mathematician worked on the basis of the rate of the present cooling of the earth. Counting backward to the time when the earth's surface must have been hotter, according to La Place's idea, he decided that our globe has been cool enough for the existence of life upon it for a period of somewhere in the neighborhood of one hundred million years. Those who try to study the rate at which mud is being deposited in our bays and at the mouth of our rivers, and who hence try to deduce how long it has taken to produce the thickness of all the stratified rock we know, arrive at a figure larger, rather than smaller, than that mentioned above. The same is true of those who try to count the age of the earth by the rate at which the present rivers are carrying away their river basins, and hence who calculate how long it has taken the rivers of the globe to wash away all the rocks which it is quite clear have been carried out. Still others have attempted to solve the problem by seeing how much salt the rivers are carrying into the sea, and consequently how long it must have taken the sea to become as salt as it is. A very late attempt has been based on the alteration in the minerals that show radio-activity. Conservative estimates, based on all of these, would give us a figure on which we must not count with any exactness, but which will serve at least to mark the present trend of opinion. We may put this figure at one hundred millions of years.

The following table gives us the names of the periods into which the geologist has divided the past history of the earth. The first column gives a simple name, which, in each case, is a translation of the technical name the geologist gives to the era. This technical name is also given in parenthesis. The second column shows the number of years ago at which this period may be placed, while the third column gives a series of names most of which are in use in geology and which are intended to indicate the stage of advancement of the higher animals in that particular period. Some of these names are perhaps giving way to later terms, but all of them will be understood by any geologist. Most of them will serve to keep very clearly before the mind of the ungeological the period which he is studying. Like all such tables, this must be read from the bottom up. This arrangement is used because the oldest rocks in the series are naturally at the bottom and the newest rocks are on the top, though occasionally a region is sufficiently upset partly to reverse the order.

TABLE OF GEOLOGICAL TIMES

ERAS	MILLIONS OF YEARS AGO (VERY UNCERTAIN)	STAGES OF ANIMAL DEVELOPMENT
Recent Life (Cenozoic)	0 to 5	Age of Man (Quaternary), Age of Mammals (Tertiary)
Middle Life (Mesozoic)	5 to 10	Age of Reptiles
Ancient Life (Pal 鑒 zoic)	10 to 25	Age of Amphibians (Carboniferous), Age of Fishes (Devonian), Age of Invertebrates (Silurian and Cambrian)
Dawn Life (Eozoic)	25 to 50	Earliest Animals and Plants

Having seen what the scientist supposes to be the method of formation of the earth itself, it will be interesting next to consider what the biologist surmises as to the origin of the life upon the earth. Here again two explanations hold. The one, and distinctly the older of the two, says that at some time in the far distant past, under conditions which are rarely if ever duplicated, out of the lifeless material of the globe were produced simple and low forms of life. These could not properly be called either animal or plant, but partook somewhat of the nature of both. Of this there is at present no evidence whatever. The only reason we have for suggesting it is that, if we understand the past conditions on the earth, there was a time when life was impossible. Now we find life. Hence it must have arisen. This of itself, of course, furnishes no proof, but leads us to try to imagine how the transition might have come about. Every scientist who believes in this form of origin holds that if the exact conditions are repeated the result will occur once more. He may believe that no such repetition is possible, but he is confident that, if it could be, life would arise again from lifeless matter.

This process of life arising from matter that is not alive is known as Spontaneous Generation. Two hundred years ago it was supposed to occur frequently. It was common belief that the beautiful pickerel weed which borders our Northern lakes, after freezing, went into a sort of protoplasmic slime out of which pickerel were produced. The eelgrass of the river was supposed to yield eels in a similar fashion. The dead bodies of animals were supposed to turn into maggots. Such crude ideas of spontaneous generation are no longer possible. The whole science of bacteriology absolutely presupposes the impossibility of spontaneous generation in the flasks and test tubes of the laboratory. One or two men of otherwise good standing in science still maintain that they are getting new life in their own test tubes, but they fail utterly to persuade the scientific world. I think it is a fair statement of the position of science to-day to say that there is no evidence whatever of spontaneous generation, excepting the presence of life upon the globe.

Not all has been said, however, on this question. The chemist is learning in the laboratory to produce many substances which, until very recent times, were produced only in the bodies of animals or plants. Dye-stuffs were originally gotten almost entirely from animal or plant material. At present the great majority of them are made in the laboratory, and in not a few cases they not only imitate the color of the older material, but actually have identically the same composition and constitution. The laboratory-made material is exactly like that made by the animals or the plants.

The same is true with regard to a large number of the fruit flavors. These are due to the presence of ethereal oils in the plant, and their exact counterparts can now be produced in the laboratory, and can serve every purpose of the fruit flavor itself. Alcohol has been produced artificially, and alcohols, which nature never dreamed of making, so far as we can tell, but which are made on her plan, are manufactured by the chemist. Last of all, sugar has recently been built up by the chemist, though the method at present is so expensive that it cannot possibly compete with the production of the commodity from the cane and the beet. As in the case of alcohol, all the sugars that nature makes can now be made artificially, and others of the same general plan which she seems not to have as yet devised can be produced within the laboratory.

Attempts have been made to manufacture proteids, but these have as yet eluded the efforts of the chemist. He is beginning, however, to come nearer understanding their composition, and when he once clearly comprehends that he may be able to reproduce them.

One of the German chemists is convinced that the nuclein in the nucleus of the cell is not a very complicated compound. Under such conditions it is not a matter of surprise that the physiological chemist should be constantly dreaming that he may at some time produce living matter in the laboratory. To the ordinary mind it scarcely seems possible. We are so entirely sure that life is not amenable to physics or chemistry that we can hardly conceive of the possibility of its originating out of matter in the test tube. If it does so come, and when it does so come, this will not prove that life is a less noble and less wonderful thing than we thought. It will only prove that chemistry and physics are more noble and more wonderful than we dreamed.

There is another way of approaching this life problem, though it seems to be rather a begging of the question than a solution of it. Of recent years it has been discovered that even the very low temperatures obtained by evaporating liquid air, say three hundred degrees below zero, Fahrenheit, do not kill seeds or spores of mold. The space between the planets is undoubtedly extremely cold. We have always supposed it to be entirely too cold for life to exist in it. But we laid little stress on the fact because we had no thought of any possible life existing there. But the discovery that seeds and spores can live uninjured through extreme cold has led to an interesting suggestion. This is that when the earth became adapted to the presence of life it was infected by germs transported on meteors from some other system. According to this theory, organic dust through space is ready to infect any planet which offers the conditions under which life may arise. Of course this theory does not explain the origin of life. It pushes back that origin a little farther or supposes that life is as old as matter itself. Again we may leave to the scientist the discussion and the elaboration of this or any other theory he may promulgate concerning the origin of life. When he has established clearly the process and can produce life we will accept his explanation; meanwhile, we will always be interested in his attempts to solve the problem, but still our simple formula, "in the beginning God," serves our present needs and will satisfy us better than any as yet unverified hypothesis.

When we find through scientific investigation how life arises we will simply know how God created it in the beginning.

The next step in the understanding of early life is to study under the microscope the simplest forms which we can find in existence to-day. This, while far easier of execution than the problems which we have thus far considered, is still not without serious difficulties. But every day brings us nearer to the understanding of the structure of living things. Life the scientist cannot see. All he can study is living matter. Whether life can exist separate from living things is a problem outside the range of his, at least present, possibilities. Therefore, concerning it he has no answer whatever to give. But when we come to study living things we find that all life is associated with protoplasm. This apparently foamy, jellylike, transparent material is the only living substance in all the world. Animals and plants are larger or smaller collections of the little masses of protoplasm which we know as cells. The lowest animals are each made up of but a single cell. This consists of a small mass of protoplasm surrounded almost always by a thicker skin or covering, known as the cell wall and enclosing a complicated kernel known as the nucleus. The protoplasm seems to be the living substance itself. The cell wall is not a simple dead scum on the outside of the protoplasm, but is itself able to do certain things which can only, so far as we know, be done by living substances. For instance, of two materials dissolved in the water in which the cell floats, the wall may permit one to soak into the animal and keep the other out. The one allowed to enter will usually be found good to be used for food by the cell. The nucleus seems to store within itself the record of its past history and thus enable the cell to do in the future what its ancestors did in the past.

Such simple cells can exhibit in very low form all the activities the higher animals show in much more elaborate development. A one-celled animal can move about, can recognize the proximity of food, can engulf its food and digest it, can build up its own substance out of the digested food, can absorb oxygen, can use this oxygen in the burning of its own substance to produce its own activities, can act in response to sensation gained from outside, can throw off its waste matter produced by its own activities, and can grow. When the proper time comes its nucleus can split in two, the cell itself enclosing the nucleus can separate into two cells, each of which can grow to the size of the parent cell and repeat its life. This is as simple an animal as we have yet discovered. Every kitchen drain swarms with such creatures. On a summer day the stagnant pools are full of them. The simplest microscope will show them clearly. This is life in its lowest terms with which we are acquainted. With such life, it seems to us, the animal and plant world must have started their existence, when first the earth began to teem with living matter.

If, then, we may form any judgment concerning the first living things upon the globe by considering the simplest creatures that live here to-day, certain facts seem clear. In the first place, life began in the water, and for a long time was only to be found in the water. Single cells are so small and dry out so easily that it is necessary to their existence that they should be kept entirely moist by the presence of water all about them. It is true many of them will stand drying, but while they are thus dried they can scarcely be said to be much more than just alive. They are utterly inactive, or, as we say, they are dormant. In such conditions they become covered with a tough skin, almost a shell, and their protoplasm is itself nearly dry. Under these circumstances the life processes hardly continue at all. The protozoa, as these small animals are called, tolerate drought for a time; but they only live, in any sense worth calling living, when water is abundant and is neither very warm nor very cold. It is safe to say that the early life of the world formed in the oceans of the time. So absolutely is the habit fixed upon cells of protoplasm that even to-day the activities of the cells of higher animals depend upon the presence of moisture. The cells of our own bodies are to-day living, as it were, in an ocean. Everyone can remember far enough back to recall some time at which a tear slipped from his own eye onto his own tongue; we know our tears are salt. The tongue has tasted, undoubtedly, the perspiration from the lip on more than one summer day; this perspiration tasted as salt as the tear itself. The lymph that constitutes the "water" of a so-called "water blister" is also salty, and even the little blood one gets into his mouth in trying nature's method of stanching the flow from a cut finger gives the impression that it contains a little salt. Every fluid of the body is salty, and every cell of the body is bathed in salt water. It is too long since the ancestors of our cells swam in the seas of the Eozoic time for us to assert with any positiveness that the ancestral habit is responsible for this trait in the descendants. Sure it is that to-day our cells, like their ancestors of old, live in water, and this water is slightly salty--as were probably the Archean seas.

The geologist tries as best he may to build up the geography of the earth in the past. He endeavors to judge from the rocks as he now finds them, where the seas, the bays, the dry land, and the mountains of earlier geological times lay. The present aspect of the earth is very recent, and earlier ages must have shown an entirely different distribution of land and water. The North American continent was certainly very much smaller than it is now. The first known lands lay close to the Atlantic seaboard and probably extended out into the water some distance beyond the present shoreline. The stretch of continent was narrow, and grew narrower as it went southward. In what is now the Canadian district, a considerable expanse probably existed in very early times. Then a great internal sea, shallower than the Atlantic, stretched its unbroken sheet over almost the entire area now occupied by the United States, while only a comparatively small hump of earth, ending in a narrower strip, lay where the great Western plateau now rears its enormous bulk.

A large portion of the history of the North American continent, with its developing animals and plants, is tied up with the gradual shrinkage of this interior sea. Slowly across the Canadian district, the Eastern and Western lands became connected with each other, while the waters progressively were pushed down the continent, which was steadily growing from the east and from the north, though less slowly from the west, into this internal sea. To-day only the Gulf of Mexico remains as evidence of the broad stretch that once extended through to the Arctic Ocean and west beyond the present position of the Rocky Mountains.

How this great Eastern backbone of the continent was produced, what sort of animals lived while these rocks were being formed, or whether this preceded entirely the existence of life upon the earth, no man to-day may surely say. In the oldest of the rocks there are beds of graphite, from which lead pencils are made. This substance is believed by the geologists to be, like coal, the remains of vegetable life. But these early rocks have been so heated and baked, so twisted and bent, that whatever forms of life they once held are now obliterated, or so altered as to give us no idea of what may have been their character.

So far as anyone can now see, this past history is wiped out forever and it will be impossible for men ever to demonstrate the character of this early life. Speculations, more or less certain, will arise. They may, after a while, seem so clear as to receive the acceptance of the scientific mind. Yet the truth remains that the early history of the earth, so far as animals and plants are concerned, is probably lost forever.

The most striking feature concerning the earliest layers of rocks in which good fossils are found abundantly is the complexity of the life. With the exception of the backboned animals, every important branch of the animal kingdom is represented, and it is just possible that we have even earlier forms of the vertebrates themselves. This, to the evolutionist, is very disconcerting. To find the great groups all well developed at least twenty-five million years ago and to find only fossils built on the same lines since almost nonplusses him. When the geologist tells him what an enormous length of time preceded the rocks in which he finds these fossils and how absolutely these earlier strata have been altered by the later geological activities he easily understands why it is impossible to find fossils in them. As a consequence, the evolutionist is forced to believe that all the earliest animals have left no clear traces behind them. Life as he first surely knows it is already extremely varied and quite well developed in some of its groups. The early animals were as well adapted to the times in which they lived as are the great majority of the animals of to-day. The reader must not infer this to mean that the animals of those days were like our present animals. They were not. No one traveling in a far country could find there animals as strange to him as would be those of the earlier stratified rocks. In these there were no fishes as we know them to-day, not a single member of the frog and salamander class, not a reptile, not a bird, not a mammal, and probably no air-living insects. It is highly doubtful whether there was any animal living upon the land and breathing the air twenty-five million years ago.

We start our study, then, at the period known as the Paleozoic era, the era of the ancient life of the globe, beginning twenty-five million and ending ten million years ago. The first of the three sections into which this period of life is divided is known as the Silurian age, the age of invertebrates. The word invertebrate is an unscientific but convenient term under which we embrace all the animals below those having backbones. This period is called the age of invertebrates because, although there is an enormous wealth of animal and plant life in the Silurian, there are no backboned animals except the lowest kinds of fishes. It was supposed for a long time that even fishes were absent. Now we know they existed, but they were small and inconspicuous. In this period corals were wonderfully abundant, particularly in the great internal sea which spread over what is now known as the Mississippi Valley. Everywhere over this region must have grown in the shallow water great numbers of creatures called crinoids or stone lilies. They were attached to the bottom by slender stems, sometimes many feet long. These stems are jointed, and when they became fossilized the sections were apt to separate, with the result that over a wide area in the Mississippi Valley it is very common to find these little segments which look not unlike checkers. At the end of the stem was a rounded head, with a mouth at the top, and around the mouth were branched, feathery arms. The creatures must have been exquisitely beautiful, but they have completely disappeared from the face of the earth, with the exception of a very few, found in the obscurity of the almost fathomless depths of the great ocean. Here they remain as peculiar relics, only preserved by the unvarying conditions in the deep sea from the extinction that has met their sisters.

Those who are familiar with our seacoast will know an interesting creature known as the horseshoe crab, or king crab, though in reality it is not a crab at all. It is rather more nearly related to the spiders than the crabs, though no one but a technical zoologist could possibly associate them together. The ancestors of these king crabs were the finest and best developed animals in this early Paleozoic time. These creatures had bodies jointed like the tail of a lobster. They were wide and flat, instead of narrow and rounded like a lobster, and each joint of the body was highest in the middle and distinctly lower at the two sides, thus forming three regions along their backs. This structure gives to these creatures the name of trilobites. These animals were the kings of the early ocean. They had an interesting habit of curling up nose to tail before they died, and, as a result, a large proportion of all the trilobite fossils we find are curled in this peculiar manner.

After these forms the most abundant fossils we find in Silurian times were creatures that at first sight looked as if they might be related to the clams. These are known as lampshells, because one shell projects beyond the other and curls up at the tip so as to resemble the clay lamps which are dug out of old Roman towns. The lampshells also have nearly disappeared in modern times. Simple creatures belonging with our present crab and snail had begun to make their appearance, but they were not as abundant as we find them later on.

The third group of the mollusks to which the nautilus and squid of to-day belong is very abundantly represented in the Silurian by fossils with coiled-up shells. As for the plant life of the time, it is exceedingly difficult to say much about it. There must have been nothing but marine plants, and these must have been on the general line of the seaweeds. Little can be definitely said concerning them.

The next period of the Paleozoic is known as the Devonian age, or the age of fishes. Now the backboned animals first make their clear and unmistakable appearance. There are remains in the Silurian which show that there must have been a few fishes at that time. The Devonian is so full of them and they are so well developed and so diversified that this period is definitely known as the "age of fishes." They do not closely resemble the fishes of to-day, but anyone would recognize most of them for what they are. Their bodies were covered, not so much with scales as with heavy plates, often arranged like tiles, those on the forward half of the animal being often larger than those surrounding the rest of the body. The creature was encased, as it were, in armor. These were the rulers of the Devonian seas. The land, as yet, was probably nearly without animal life, the creatures thus far being almost confined to the water. A few insects make their appearance and a few thousand-leggers are running around among the lowly plants; a few spider-like animals have arisen; there are a few snails that have left the water and taken to the land. Altogether only the dawn of a land fauna is to be noticed. In the Devonian the plants are creeping up upon the ground. Ferns are growing about everywhere, though they are not exactly our ferns, but are rather a sort of intermediate form between these and the present seed plants.

Now comes an entire change in the history of the world. By some means a rise in the bottom seems to have cut off a great part of the internal sea from the outer ocean and to have converted it into a widespread shallow bay, much like the sounds which lie back of the islands that line the Atlantic Coast from New Jersey to Florida. Just as this coastal region to-day is covered with salt marshes, so the whole internal sea of the Carboniferous period was converted into a great swamp. Sometimes an oscillation of the crust of the earth brought this marsh above the surface of the sea and a luxuriant growth of plants spread over it. Then a sinking of the bottom allowed the mud and sand to wash down the shores, and spread out over the marsh, and enclose the muck of the marsh under a layer of sand or clay. Another lift of the bottom would start the swamp growing once more, and a series of alternations between marsh land and sound seems to have followed. The plants of this period are not the plants of to-day, though we still have some very degenerate representatives of them. The common horse-tail, with its angular, slender, leaflike branches and its club-shaped spore-bearing body, is a modern degenerate descendant of the treelike calamites of the Carboniferous forest. A creeping evergreen, known by the name of clubmoss, is in like manner the modern degenerate remnant of the scalestem and sealstem, which were the great trees of the forests of the coal period.

All over the surface of the marsh, between these big trees, grew the ferns. While the coal itself was formed generally from the scalestems and sealstems, the most common fossils found in the shales that lie upon the coal beds are the ferns which covered the surface of the marsh.

It is believed by many geologists that this great luxuriant forest points to a time when the climate was far warmer than it is to-day, when the air was moist and heavily laden with carbon dioxide, and when a great mass of clouds practically enveloped the earth. In this way only do most geologists account for the enormous wealth of vegetation in the Carboniferous period and for the abundance of plants up to the Arctic Ocean, of the kinds that now grow chiefly in the tropics. But of recent years a few geologists point to the fact that the peat bogs of to-day, which seem to be the beginnings of future coal deposits, are found almost entirely in cold countries. Hence it is a serious matter to attempt to describe the climate of any part of the Pal 鏊 zoic era. Certainly of the climate earlier than the Carboniferous it is very risky to say anything definite.

The forests of the coal period seem actually to have cleared the air; at least now we begin to find creatures related to our salamanders and frogs moving about among the stumps of the marshes. These amphibians are evidently the descendants of some of the fishes of the Devonian times. Among these fishes were some which bear a great resemblance to a few found in South America, in Africa and Australia to-day, and which we know as lungfish. Anyone who has cleaned our fresh water fishes in preparation for the table will remember that inside of them there is a long slender bladder filled with air. This bladder assists in making the fish light, hence making it easier for it to support itself in the water. In certain swampy regions these lungfish swim freely in the water of the marshes. When the dry season comes, however, the water evaporates, draining the marshes completely. This would prove the death of most fishes. The lungfish have a curious habit which keeps them over the dry season. They cover themselves with a coat of mud, inside of which there is a lining of slime produced from their bodies. In such cocoon-like cases they survive the drought. The means by which they breathe during this dry season is interesting. The swim-bladder which we have just described in other fishes is, with this lungfish, peculiarly spongy in its walls, presenting a large surface full of blood vessels which absorb the air on the inside of the bladder. This air the fish changes with moderate frequency, the result being that the swim-bladder serves him exactly as the lung serves a higher animal. To this fact he owes his name of lungfish.

We sometimes gain much light concerning the past history of any particular form of animal by studying the development of that animal in the egg, or, in the case of the mammals, before birth. It is an interesting fact that when the lung begins to form in the embryo it starts as a simple sac which is an offspring from the gullet, and occupies the position of the swim-bladder of the fish. This sac later divides into two, and develops into the lungs of the animal. This assures the zo 鰈 ogist that the origin of the lungs in the higher animals is found in the swim-bladder of the so-called lungfish. In this Silurian time certain of these lungfish were perhaps trapped in the basin in the marsh by the uplifting of the border. The waters becoming progressively shallower and more crowded, these fishes took to the land, their fins developing into awkward limbs which slowly became more perfect.

To state the fact in this simple fashion is to make it seem far less probable than is really the case. The simple forms of the life of lowly creatures, as well as the simple character of the legs and feet in the salamander class, make the explanation not so unlikely as would at first sight appear. Suffice it to say that the scientist now believes that out of the lungfish of the Devonian came the amphibians of the Carboniferous period.

At the end of the coal period came the greatest change the face of the globe had seen for many millions of years. Slowly the continent rose on both sides of the old interior sea. A great plateau formed in the region of the Alleghenies and another in the western district, though this latter uplift was to be completely washed away, and later to rise again into the Rocky Mountains and the Sierras. With the uplift at the edges of the continent came a steady rise of the internal marshes, until what had previously been swamp land became progressively first dry land and, in the western part, even desert, in that respect being somewhat like what it is now.

The amphibians of to-day (animals like the salamander and frog) all lay their eggs in the water and their young have a tadpole stage. This doubtless was true of the amphibians of the coal period. With the beginning of the Mesozoic, or "middle life" period, a change and a progression comes over the animal world. The tadpole life of the frog is a rather lengthened one, while the toad has learned to crowd its tadpole life within a few weeks. It would seem as if, in the earlier times of the Mesozoic, this same change of habit had been going on. With the drying up of the swamp, some of the amphibians crowded their tadpole stage further and further back, until it was completely accomplished before their young left the egg. An examination of the development of the reptile in the egg will show a stage very similar to the fish and to the amphibians, but this is all experienced before the reptile emerges from the egg. The reptilian egg, unlike that of the frog, is covered with a shell, packed away under the surface of the ground, and left to its own fate. If, as most geologists believe, the climate of the Mesozoic was distinctly warm, this habit of the parent of forsaking the egg was not a serious matter. However the creatures arose, it is certain that in this Mesozoic age reptiles roamed the forests, swam the seas, and even flew in the air. Probably at no other time in the earth's history has any one class of animals so completely dominated the situation as did the reptiles of this age. They were not only abundant; they were frequently enormously large. Their skeletons are among the most interesting that we find to-day. Gigantic lizards, seventy feet long and eighteen feet high at the shoulders, dragged their heavy bodies through the marshy edges of the lakes. Out upon the land others, not quite so heavy nor so large, roamed about, some of them feeding upon the soft vegetation, others having teeth fitted to tear down their herbivorous cousins. In some of them the hind legs and tail were very heavy and the front legs so light that it is quite clear they must have hopped around as do the kangaroos to-day. Others of these reptiles went back to the sea, lost the leglike development of their limbs and regained the flipper form, though the bones of the fingers and toes are singularly distinguishable in the paddle.

Strangest of all, a considerable group of these wonderful reptiles lengthened their little fingers, sometimes to three or four feet in length, and had a skin stretched from these fingers over to the body in such a fashion as to give them wings not unlike those of the bat. In the wing of the bat, however, four of the fingers of the hand run through the membrane and support it. In the pterodactyl, as these flying reptiles are called, the middle finger supports the web, while the remaining fingers can still be used to clasp objects or serve the animal to lift himself, as the bat can do with his thumbs.

Meanwhile an entire change is coming over the plant world. The last third of this age of reptiles is known as the Cretaceous or chalk period. Now, for the first time, the forests begin to take on more of the character of our forests of to-day. Plants like our willow and beech, poplar and sassafras appear in great abundance. Their broad leaves serve better than those of any earlier plants to catch the sunlight. But in addition they offered such effective evaporating surface that they cast off rapidly the moisture obtained from the ground by the plant. Accordingly in the winter season, when the water in the ground is frozen and not available for plant purposes, they were forced to throw away their leaves. It is quite possible that up to and including the time of the Carboniferous, plants were all evergreen. There had been before this little variation in climate over the globe. Life in the Cretaceous begins to take on distinctly its modern form.

Among the reptiles of the forest there appear to have been a few small creatures which to an observer of those times, if there could have been an observer, would have seemed of the utmost insignificance compared with their giant cousins.

These little creatures climbed up into the trees to escape their enemies. There were some in whom the skin, in front of the elbow and behind the wrist, was loose, and stretched across the joint a little like the wing of a bat. This reptile, climbing into the trees to escape its enemies, found that this loose flap of skin served it nicely, and sailed out of the trees in a manner not unlike that of the flying squirrel of to-day. Among these experimenters in aviation, certain forms produced scales which became elongated and finally slit up along the side. These slit scales slowly developed into the feathers of the birds of to-day. Whether the steps by which the change occurred have been correctly stated or not, the result is sure. In the rocks of the chalk period we find the remains of an interesting creature. If nothing but its bones had been found it would have been called a reptile. It had a long tail, it had claws on its front limbs; it had teeth in its mouth; it had a flexible backbone. All of these are reptilian rather than bird characters. Yet on the rocks surrounding these bones are the unmistakable impressions of the feathers of the wings and of the tail. Nothing in the world to-day has feathers excepting the birds, and in this "ancient winged thing," for this is the significance of its name-- arch 鑿 pteryx--we have perhaps the most remarkable link in the world between two distinct sections of the animal kingdom. Here is a creature half reptile, half bird; perhaps one-third reptile and two-thirds bird. It was about the size of the crow. A little later unmistakable bird skeletons will appear, but still their jaws are provided with long conical teeth.

Still more interesting from our standpoint is another set of primitive animals, utterly insignificant in appearance, but of momentous importance on account of their later history. Among these reptiles were a few small creatures perhaps not much bigger than mice or moles. Their teeth were a little more complicated and specialized than the teeth of their reptilian cousins. Between their scales were small and sparse hairs. Almost nothing but their jaws remain to-day to tell us anything about them. But in this humble little creature of the Mesozoic, utterly insignificant beside the tremendous reptiles of the time, we discern the ancestor of the mammals. These were the progenitors of the horses and cows, of the cats and dogs, of the monkeys and apes, of the men of to-day.

During this chalk period, which forms the last portion of the age of reptiles, life for the first time grew to look much as it does to-day. Now, apparently, the cold of winter and the heat of summer followed each other in regular succession. There have been colder and warmer periods at various times in the previous history of the earth, but undoubtedly they were more uniformly cold or uniformly warm than now. Ages were warm, or ages were cold, but now the earth clearly shows the annual alternations of summer and winter, and for the first time clearly shows the bands of climate on the earth which we know as zones.

In the chalk period this new factor of cold works mightily in favor of the mammals. Their reptilian ancestors were cold blooded. When the climate was warm they were active; when the climate was cold they were sluggish. With the continuation of the annual alternations of cold and warm weather that had now set in upon the earth, the little birds and mammals had in their warm blood an advantage which, in the long run, enables them not simply to compete with their reptile forefathers, but to outdistance them absolutely in the race. Here and there, on earth to-day, exist a few big reptiles like the crocodiles and the boa constrictors. But they are few and comparatively insignificant among the multitudinous population of the globe and are confined to the hotter portions of the earth. For the most part, the reptiles now play an insignificant and unobtrusive part. The little molelike creatures, practically unnoticed between their feet in the later Mesozoic, have come to supplant them entirely, and almost to rival them in size. While the reptiles have grown steadily smaller, the mammals have steadily become larger.

While there is no land mammal to-day as big as the heaviest of the reptiles in the Mesozoic, the whale, which is one of the mammals that has again taken to the ocean, surpasses in size even those gigantic creatures. There never lived in the world before a creature quite so big as the biggest of our whales. Size, however, is not the most important point in any animal. Speed, sagacity, variability, and power of adaptation, these are the qualities which the world prizes, and these the new mammals possessed.

The next geological era is the Cenozoic, or period of modern life. This is divided into two quite distinct sections, the Tertiary and the Quaternary. This era began about five million years ago, roughly speaking, and is still going on. The greater half of it is known as the Tertiary. It was during this time that the mammals came to their own. At first these creatures belonged to what the scientist knows as generalized types. They are jacks-of-all-trades. The student of early animal life finds in the little Phenacodus, which was scarcely bigger than a good-sized setter dog, the beginnings from which many forms have subsequently developed. This creature showed points of structure which to-day may be seen in such diversified animals as the dog, the horse, the rabbit, and the monkey. It is not, of course, suggested that Phenacodus was the immediate ancestor of any of these. But there were no animals in those times more like these I have mentioned than was Phenacodus, and from forms like it in main features all of these other animals have since been derived, each species of animal having become adapted to one particular kind of life. The development of diversified situations on the earth, the varieties of climate, the variation between marsh and upland, between valley and plateau, furnish a complexity of environment into each niche of which a new form of animal fitted itself.

With the increased complexity of mammals comes the submergence of the reptiles and amphibians to-day. In all sorts of situations we find mammals. The old-fashioned continent of Australia is separated from everything about it by deep water, impassable to any animal which lives upon it. In this secluded country evolution is very slow and animals are very antiquated. We still find there mammals with the ancient habit of laying eggs in a hollow in the ground, though after these eggs are hatched the young are nursed on the milk of the mother. But on the great continental stretches, where competition is keen, where the animal must battle for his life against a wide field of other animals, where migration into new situations is possible, the rapidity of the development has been very much greater.

It is in such a situation that man has arisen. In the extreme southeastern portion of Asia, and on the islands lying close to the coast, his highest non-human relatives, members of the ape family, have reached their best development. These, of course, are not man's ancestors. They are the less progressive members who are left behind entirely in the race. Whether we have to-day any traces of the steps by which man arose from the animal beneath him is vigorously disputed. Eminent scientists will be found on both sides of this question.

Many scientific writers to-day take it for granted that one form, discovered in Java, while it may not be in the absolutely direct line, must be very close indeed to the line of ascent toward man out of the apelike forms. A scientist by the name of DuBois, working in the banks of a stream in south-central Java, found a thigh bone which seemed to him exceedingly human in its general character and yet not absolutely like the human thigh bone. The oncoming of the rainy season raised the water in the river so that DuBois could not continue his search. Returning a year later, and digging back deeper into this bank, he found a skull cap and two molar teeth which seemed to him to belong to the thigh bone, although they lay several yards farther back, but at the same level in the bank.

When these bones were subsequently presented to a meeting of European scientists by DuBois, he claimed to have found the "missing link" for which there was so eager a demand. Some of the best anatomists of the meeting, notably Virchow, laughed at his claim and said that the skull cap was simply that of a human idiot, and could be duplicated in any large asylum. A committee of twelve naturalists was appointed to report upon DuBois' find. Of this committee three asserted the bones to be those of a low-grade man, three insisted that they belonged to a high ape, of a type somewhat higher than any we know to-day, but still distinctly an ape. Six members of the committee of twelve agreed that the remains were those of a creature higher than an ape and lower than any normal man, and represented, in their opinion, a stage distinctly along the line of development out of the apes and into man.

This so-called "Java find" is known in science by the name of Pithecanthropus, which means the ape-man. Whether we look upon this fossil as a serious find or not, it is very certain that in the caves of Europe belonging to the Quaternary period we find abundant evidences of primitive man. The older these evidences are, the more likely they are to be distinctly below the grade of man of to-day, in the size and shape of the brain case and in the length and massiveness of the jaw.

There are probably more races than one represented among these skulls. Some of them are surely well-deserving of the title of low brow. Their heavy ridges over the eyes, their small foreheads, their massive, heavy-set jaws show a race of men far less endowed mentally and much better endowed in the matter of brute force than the men of to-day. These skeletons, or parts of skeletons, are turning up every year, and we are just beginning to know much about them. Capable men are studying them with much care. The next fifty years may not improbably make the history of the ascent of man as clear as is now that of the horse, to which we shall refer later.

The whole question of the descent of man from the lower animals, or his ascent from them, as Drummond aptly termed it, is to most people so entirely repugnant as to set them at once, and finally, against all willingness to consider the question of Evolution. This, however, does not solve the problem. Even though truth be horribly unpalatable, it is still to be believed if it is only the truth. There is practically no doubt left among scientific men of the origin of man in lower forms. The evidences grow more and more complete year by year, and from every line of investigation. Whether we study his anatomy, his embryology, his history, his language, or his civilization, all indications point in the same direction. Constant discoveries indicate the fact of an enormously long development from a very humble form. If this proves to be true and remains unpalatable, the fault lies in the palate and not in the truth. Gradually we are coming to understand that there is no reason why this truth should be unpalatable. We consider a rise from humble conditions to be the glory of our heroes; we esteem it an added charm in their strength that they should have developed from untoward surroundings. It is not a disgrace to man to have descended from the apes. It is to the glory of man that he should have ascended from forms not much more promising-looking than the apes of to-day. We must repeat, however, that the apes were the unprogressive members, and hence we must not judge man's ancestors too harshly. It must have been in them to rise. But the great glory in the thought of the humble ancestry lies in the possibilities of his future. If out of a creature not materially unlike the gibbering ape of to-day there should have come, under the guiding hand of an Almighty God, creatures with the endowments and capabilities of man of to-day, then this is only an earnest and foretaste of that which may be expected in the future. A time will come when man shall have risen to heights as far above anything he now is as to-day he stands above the ape. Even then there seems no end. With Infinite Power as the agent, and limitless time in which to work, man would be limiting God to an extent unwarranted by the history of the past to imagine that His process had stopped to-day, and that man, with his many imperfections of body, of mind, and of morals, should be the best that is yet to come. There cling to him still the limitations and dregs of his brute life. Often the brute in him comes to the surface. Little by little he is coming to be dominated by the qualities God has last given him. Slowly the brute shall sink away, slowly the divine in him shall advance, until such heights are attained as we to-day can scarcely imagine. As we can scarcely conceive the beginnings of this process, so we can with difficulty imagine its end. This only can be seen by the Eternal through whom it shall all come to pass, and by whom all will in time be accomplished.

CHAPTER VII

HOW THE MAMMALS DEVELOPED

When the idea of evolution first began to be much discussed, especially after the publication of the "Origin of Species," there were several points which appeared to be more than commonly difficult of explanation. It did not seem impossible that the various types of domesticated cattle should have descended from a common ancestor. It did not seem difficult of comprehension that the dog might once have been a wolf. Though not quite so credible, it did not seem absurd that the tigers, lions, and leopards should have once all been alike. The resemblance between these are strong enough to make the idea seem conceivable. Though men were willing to concede this much, they insisted that the great branches of the animal kingdom varied so widely from each other as to make it certain that each was a separate creation. It was particularly objected that the mammals differed so entirely from other animals in several important particulars that a special divine act was necessary for their appearance. The mammals have a furry covering entirely different from the clothing of any other animal in the kingdom, and have warm blood, which is found nowhere else except among the birds. But particularly their method of producing their young seemed so entirely different from that of any other group that here a special creation was deemed absolutely necessary.

Other young creatures are produced from eggs laid by the parent and subsequently hatched. The young of the mammals are born alive and comparatively well developed. In addition, their first food, the milk of the mother, is so entirely different from the food of any other creature that this again seemed to involve a separate creation. Gradually we have come to understand the whole matter of reproduction very much better. Minute and careful dissections of rabbits, of dogs and cats, of animals slaughtered for food, with occasional post-mortem examinations of human beings in various stages of the development of the young, leave us no longer in doubt concerning the main features of the process. The better we come to understand it the more clearly it becomes evident that in the development of the mammals we have no new procedure, but, as in so many other activities, new developments of an old process.

There are two entirely different methods by which new animals and plants may arise. One sees sometimes in the home of a friend a geranium of particular beauty, the like of which he would be glad to possess. The accommodating friend cuts a small piece from the geranium. This is stuck into poor but well-watered ground, develops roots, and eventually grows into a geranium stalk exactly like the one from which it came and of which it is in reality only a detached part.

In similar fashion, if one wants a particular kind of apple, he never trusts to planting an apple seed. Going to the tree of the variety he desires, he takes from it a small twig provided with a bud and inserts this bud into a cleft made in the young branch of another apple tree. The young bud so inserted starts up into a new branch, resembling almost absolutely, not the tree which feeds it with sap, but the tree from which the bud was originally taken.

When we wish a particular variety of potato we obtain pieces of the potato of the kind we desire. Each of these must contain an eye, which is a bud of the old potato. When the sprout appears the new plant will be practically identical in character with the plant from which the potato was taken. This sort of reproduction, in which a piece of the old parent grows up into the new generation, is called the asexual method. But one parent is concerned in the process, and the offspring are as nearly as may be like the parent from which they arose.

The gardener who wishes to obtain new varieties is not content with this method. If he plant the seed of the potato the outcome will be most uncertain. His seed must be taken, of course, from the fruit of the potato, and most of these plants never fruit. Every grower of large quantities of potatoes will have noticed occasionally, on the tops of the plant, after the flowers disappear, a globular growth looking not unlike a small tomato, but with a tendency to become purplish green in color. This is the fruit of the potato and in it are the seeds. When these are planted all sorts of potatoes are liable to start up. Most of them will prove worthless. An occasional seed may produce an uncommonly fine plant. This new variety may thereafter be propagated from the tuber, as the potato itself is called, and the new strain will be kept constant in this way. This method of using the seed for reproducing the plant is called the sexual method, because two parents cooperate in the production of the seed. The pollen came from one parent and the ovule, which after fertilization swelled up into the seed, came from another. By this combination of two individuals new varieties become quite possible. Nature seems to be more concerned in improving her strain than in maintaining her older strains. In all of her lowest plants and animals she uses the asexual method of reproduction. As we go higher in the organic world the two-parent method becomes increasingly common. When we reach the higher animals, and most of the higher plants, this plan of double parenthood, the sexual method, alone is used.

In order that we may the more clearly understand how the mammals produce their young and nourish them, we shall begin at the lowest class of the backboned animals and note how the process is there accomplished. As we pass upward through the kingdom the method acquires greater complexity. When we finally reach the mammals, what at first seemed an absolutely new process will prove to be, as is all of nature's work with which we are thoroughly acquainted, but a modification and an elaboration of some previously existing process.

Some time ago I was passing the early months of summer by the side of a lake in northern Pennsylvania. Near my tent, on the edge of the water, was a wharf from which it was possible to look down into the shallows about the edge of the lake. In early July the bottom began to take on a strange appearance. Spots as big as a dinner plate became evident because they were cleaned of the finer sand or mud which is common on the bottom. A close examination showed that each of these circular spots was being occupied and cleaned up by a sunfish. The pebbles were lifted into the mouth of the fish and driven out again with force. The water which emerged with the stones seemed to wash away the dirt, while the pebbles themselves became gradually cleaned of the green plant life which ordinarily covers them. After the process was completed each spot was saucer-shaped and free from scum and mud. Over each of these spots hovered the sunfish which made it, and round and round the fish swam. The circles thus traversed were so near each other that every now and then the occupants of two adjoining nests would meet on the border. The fish which was most nearly on its own ground would at once attack the other and drive him away. In a few days the other partner in each family seemed to appear. Now two fishes swam side by side over each nest, bringing the lower edge of their bodies comparatively close together. In this position they moved around over the pebbly bottom. The female was discharging her multitudinous and very small eggs, so that they dropped to the bottom of the nest. At the same time the male was expelling what in fish is known as the milt. In this milt are the sperm cells of the male, each consisting of a rounded head and a very slender body. These are attracted by the eggs. Pushing up against them, the head of a sperm cell, consisting almost entirely of the nucleus of the cell and carrying the determinants which were to decide one-half of its future characters, penetrated this egg and fused with its nucleus. This was filled with the determinants of the characters inherited from the mother. Of course many of the eggs, of which probably there are a thousand, must have escaped fertilization. There are doubtless a thousand sperm cells that went to utter waste for one which found an egg to fertilize. These eggs nestled in the crevices between the stones in the warm water of the edge of the lake. Here the sun could easily penetrate to the bottom and hatch them. The little fish, still guarded by one hovering parent, swam around in the water long before the yolk of the egg, containing its large amount of food, had been absorbed into the tissues of the young fish. This fatty store made the abdomen of the fish in which it lay protrude enormously. Gradually the fish grew larger and the yolk grew smaller until all had been consumed. Soon the fish began to forage for himself and no longer to demand or care for the company and protection of its parent. The little sunfish is highly favored among his comrades in having any care whatever by the parent. In the case of most fishes the female, swimming slowly over the bottom, deposits her eggs, which are fertilized by the male, which follows behind her. After the eggs have thus been laid and quickened no other attention is paid to them by either of the parents.

Fish are stupid almost beyond the comprehension of those who are not students of the minds of animals. Frogs and toads are a distinct step in advance, and hence their mental activities play a larger part in the process.

In the love-making of the frogs and toads the song has an important share.

In each species the voice is a little different from that of any other. In our familiar garden toad we have an excellent illustration of the method common to the entire group. When spring comes an impulse seems to stir in all the toads of a neighborhood. Heretofore they have stuck faithfully to dry ground; now they start off for the water. Whether their impulse is simply to move down hill or whether they by some means detect the near presence of water, I cannot say. Certainly a new fountain on a lawn will secure in spring its prompt and full share of the neighborhood's toads. In any event the toads of a district congregate in great numbers in any pond or along the edge of any moderate stream. Within a short time their flutelike, quivering voice is heard far and wide. That this note has an attractive power over the female there is no doubt. She herself makes no effort to imitate, but the song of her mate is persistent and exceedingly sweet. I have seen a male sit upon a clump of grass and utter his love call. Before he had been singing for more than half a minute three females hastened toward him from a distance of perhaps twenty feet. Each seemed anxious to reach as promptly as possible the creature whose voice had proved so attractive. When the mating comes, the female discharges a series of small shotlike eggs which are encased in a very tenacious mucous. While they are being deposited the male fertilizes them. No sooner have the eggs, fertilized by the sperm cells, reached the water than the mucous at once begins to swell. The result is that eggs appear encased in two slender strings of jelly, each having a diameter about that of a lead pencil. At intervals of not more than half an inch the shotlike eggs may be seen. The mother toad, in laying these eggs, moves about rather restlessly in the water. By this means she succeeds in wrapping the strings about the grass and sticks of the pool. This will hold them quite safely even against a considerable current of water, should the stream rise and flood the side pools in which the eggs are laid. With this amount of care, however, the attention of both parents to the young entirely ceases. They are now abandoned to the chances of a fortune to them exceedingly unkind. A toad will lay about five hundred eggs. It is evident that on the average only two of these can attain maturity by the time the parents have died, for the number of toads does not materially alter season by season. The connecting string is made up not of nourishment for the eggs, but of a bitter mucous so unpleasant to the taste that fish are thus deterred from eating the otherwise nourishing material. This secures for the young embryo a chance to mature which in the absence of the jelly it would entirely lack. Imbedded in this mucous is the embryo itself, surrounded by a small amount of albumen and containing inside of itself a very considerable amount of yolk. This gives to the egg a volume possibly a hundred times that of the egg of the sunfish. Thus, even counting the care the parent sunfish took of its offspring, which care is very uncommon among fishes, the toad stands a distinctly better chance in life. The protection of the bitter mucous and the large amount of yolk permitting considerably larger development before leaving the egg, give to the toad a material advantage. When the toad first emerges from the egg it is amazingly like the fish. It has gills at the side of its neck and swims by the movement of its tail. Later its limbs develop, the hind ones coming first, its tail is absorbed, and it is now a true toad, ready to leave the water.

Altogether a higher state of reproduction is encountered when we reach the reptiles, which are the next higher class of backboned animals. Here very distinct developments of the process are discovered. The turtle, to use the best known illustration, may lay but twenty eggs. But she will not lay them at random in the water, as do the toads and the fish. Each egg is wonderfully fattened with yolk. This means that it is possible for the creature to develop to a far greater extent before leaving the egg than was possible in the case of the toad. Accordingly the little turtle, while it begins life not unlike a fish and goes through the gilled and tailed period, during which it is not unlike a tadpole, passes beyond this period before leaving the shell and has already acquired its full turtle characters when first it steps upon the scene. So big an egg as this would be highly nutritious and animals would desire it immensely for food. Hence it becomes necessary for the turtle to securely hide her eggs. In order to do this, she scoops out a pit in the sand in which she deposits them and here they develop. If no further provisions were made the eggs of the turtle would dry completely and never hatch. Accordingly it becomes necessary for the turtle to enclose each egg in a tough, leathery membrane, known as the shell. Because the egg is thus encased it is necessary for it to be fertilized before being laid. Accordingly the male must place the sperm cells within the body of the female. These cells swim nearly to the top of the tubes in which they are placed, and there fertilize the descending eggs. Farther down the canal the shell is secreted about the now swollen mass of yolk and white, completing the egg just before it leaves the parent.

If the evolutionist understands properly the line of descent, the birds and mammals are both the descendants of the reptiles. While there is less exterior resemblance between a chicken and a turtle than between a cat and a turtle, the real relationship in the first case is much closer than in the second. This is perhaps most easily seen in the scaly legs of both bird and reptile. Another remarkable resemblance lies in the fact that in both cases the eggs are large, well stored with nourishment, and protected by a resistant shell.

So few people know the turtle's egg that it will be better to describe that of the hen, which it largely resembles. Underneath the hard shell is a tough but flexible membrane which lies against the limey coating, except at the blunt end, where a separation between the two gives room for a bubble of air. Inside of this shell and its membrane lies the white of the egg, which is nourishment for the chick, though not nearly so rich as the yolk. This, besides the albumen which it contains, is stored with large quantities of fat. It will be remembered that upon breaking a hen's egg and dropping it into a bowl, the yolk holds together because it is enclosed in a delicate sac. As the yolk falls into the bowl there floats to the top of it a lighter yellow spot as big as the end of a lead pencil. This is all of the egg which thus far represents the chick itself. All the rest is nourishment. This disk already consists of three reasonably distinguishable layers of cells, which grow rapidly different from each other. They spread and bend and twist, forming the young chick and a set of organs which serve for its protection and maintenance during its embryonic life. Within a few days these accessory organs will have formed distinctly. Within the upper half of the yolk will be found the small developing chick, which for the first thirty-six hours of its development passes through a stage not unlike the fish, or the earlier steps of the turtle. Within a few days it becomes clearly evident that this creature is to be a bird, though it is much longer before it is clearly a chick.

This embryo is so soft that it is almost like curd in thickened milk, and could be very easily destroyed were it not for a protective device which Nature has employed. It seems necessary that it should be protected with the utmost care. The matter will be better understood if we recall a common experience. Almost everyone has tried to dissolve some substance in water in a vial. If the bottle be filled with fluid to the top and corked it is very difficult to shake up the contents. Even vigorous agitation produces little movement of the material on the inside. If we wish to shake up the solid with water the bottle must be left partly empty. The brain of a human being is protected by just the same device. If it simply lay within the skull the first fall would mash the gray substance against the side of the cavity. To prevent this calamity the bony case is made somewhat larger in capacity than the brain itself, and the space between the two is filled with a watery fluid. This serves to prevent jars and shocks. In the hen's egg the same plan is pursued. The embryo lies on the inside of a bag considerably larger than itself. This sac, called the amnion, is filled with a watery fluid. With such a protection only the most severe shock to the egg would sufficiently jar the embryo to do it any harm. The ordinary experiences of an egg leave it undisturbed.

Every living creature requires a constant supply of food and of oxygen. The embryo is a living creature, and is no exception to the rule. It needs an abundant supply of easily assimilated food and of oxygen. When the hen's egg is first laid the entire contents, with the exception of the little light-colored disk which floats on the top of the yolk, form the nourishment. The disk alone is the living organism. In the earliest stages the embryo receives its food by simple absorption from the yolk. As the chick increases in complexity the yolk at first grows swampy, with fluid trickling here and there through the more solid portions. Thin walls form about these little streams, thus producing blood vessels which cover the entire surface of the yolk. These absorb the nourishment and turn it over to the embryo. As the latter grows in size both the yolk and white diminish. The embryo soon becomes larger than the remaining yolk and is attached to it by a cord filled with blood vessels which enter the chick near the center of its body. The abdominal wall has an opening at this point. One of the later occurrences in the life of the chick, before it breaks through the egg, is to have the last remnant of the yolk and its sac slip to the inside of the abdomen, which then completely closes over it.

As yet, we have seen no arrangement for furnishing air to the chick. At the same point at which the blood vessels from the yolk enter the chick, another set of vessels pass in and out. These are attached to a large flattened bag which floats above the embryo against the upper side of the shell. This bag is called the allantois, and serves as a sort of lung for the developing chick. The shell is porous enough to allow air to pass through it. The blood vessels of the allantois take in oxygen and give out carbon dioxide through the porous shell. The blood thus altered is returned to the chick and serves its life purposes. One of the reasons why the chicken must turn its eggs in the nest is that, if the allantois remain too long in contact with the upper shell of the egg, it will become attached to it and will not thereafter perform its functions.

The embryo thus enclosed in the egg finds its protection in the fact that it is encased in a fluid contained in the amnion. It draws its nourishment from the yolk upon which it lives and the nourishment is transmitted to it by blood vessels. It draws its oxygen and throws off its wastes through the instrumentality of the allantois, which covers it over. Day by day the chick becomes larger, day by day it grows to look more like what it is to be. By the nineteenth day it appears to be complete. Its nervous organization is, however, not thoroughly developed. If removed from the shell the chick still is indisposed to stand upon its feet or to run about. If allowed to remain in the egg until the twenty-first day, the chick will be able to push its beak through the skin enclosing the bubble of air at the blunt end of the egg and get the first breath into its lungs. Now it gives a faint peep, breaks the shell of the egg, and steps out into the open air.

I have given this somewhat lengthened description of the development of the chick because of the light it throws upon the method pursued by the mammals. The features which have been described in the case of the chicken's egg could be as fully observed in the case of the turtle or any of the other reptiles. Mammals are descended from the reptiles of the Mesozoic, and whatever peculiarities there may be in their method of producing their young must be derived from the reptiles. If we wish to know how the earliest mammals produced their young, we can only judge by the lowliest members of the group that live upon the earth to-day. The most primitive of these is the so-called Duckmole, of Australia. This little creature has habits not unlike those of the muskrat. It burrows in the bank of a stream, and makes a nest at the end of the burrow, where it lays its eggs. This is one of the very few warm-blooded, hair-covered animals which still lays eggs. A little higher in the scale stand the kangaroo and the opossum. These creatures keep the egg inside of the body until it is hatched. But this happens in so short a time that the young animal is exceedingly immature and as yet unable to stand the outside air. Accordingly there is a double fold of skin on the abdomen of the mother, covering her breasts. This forms a suitable resting place into which these young are conveyed as soon as they are born and from which they do not emerge for many days. The little creature instantly fastens upon the nipple of the mother, keeping its mouth constantly in this position. At intervals the muscles of the breast force the milk into the mouth of the young, which is still too undeveloped to suck for itself. As it gets older the little opossum or kangaroo emerges from the pouch in the pleasanter part of the day and in the absence of danger. It returns to the mother's pocket as soon as it becomes cold or a cry from its parent warns it of its defenseless position.

These creatures are the lowliest of the class upon the earth. The great majority of all mammals have elaborated a far finer plan, in which the young are retained within the body of the parent until they are quite able to stand the air. The length of this time varies in different mammals from a few weeks to more than a year. The egg must be fertilized before it leaves the body of the parent. If it should fail in this it simply passes out and is wasted. If the fertilizing cell reaches the egg before it has progressed far down the tube it begins its development. The embryo forms for itself the sort of head and tail and gill slits which would have served its fish or its tadpole ancestor. Its limbs develop as little buds indistinguishable from similar buds that would have formed fins for the fish or wings for the bird.

Around the embryo there forms a sac, the amnion filled with a fluid which serves to protect the young mammals exactly as the growing chick was protected. Under the forming creature there hangs a small but empty yolksac. This is an actual remnant, a reminder of the past, when the eggs of the mammals were also packed with yolk and the growing embryo secured its nourishment exactly as does the maturing chick. But a new method has been provided for the mammal, and consequently the yolksac, though it has not entirely disappeared, has no nutritive content for the growth of the embryo.

The allantois of the chick now gains a new development and an altered function. In the case of the chick it floats against the shell of the egg and absorbs oxygen through the shell. Inside the body of the mammal this is impossible, because the air is too far away. No shell is formed about the egg because it is not to be laid. The tube of the parent's body in which the egg lies becomes thickened at the point of contact with the egg. It grows spongy and full of blood vessels. Meanwhile the allantois is also growing spongy. These two tissues are so closely pressed against each other that the blood vessels of the transformed allantois mesh in with those of the thickened parent wall. Thus the blood vessels of the mother are brought into close contact with those of her offspring. Her blood seeps over into the transformed allantois which is now called a placenta. From this it is handed over to the offspring, which thus receives from the mother her blood, and returns its own used blood for enrichment and purification. So the allantois of the reptile has become the placenta of the mammal. In the first instance it served only as an organ of respiration. Now it has come to supply the embryo with rich blood containing both food and oxygen derived from the mother. After the offspring is born this thickened pad breaks loose, and subsequently is also extruded from the body, forming what is known as the afterbirth.

Thus far we have spoken of the change in the method by which the young are brought to such a stage of development that they can stand the outer air. One of the improved differences between the mammals and other animals lies in the method by which they nourish their young for some time after birth. The very word mammals signifies an animal who is in the true sense of the word a mamma. This name for mother is given to her because of the fact that she possesses what are technically known as mammary glands, or, in simpler language, breasts. It would seem as if here we had an entirely new organ. No other animal gives nourishment to its young in such fashion; all mammals do. What is the origin of the habit? How did the organ arise?

A part of an animal's body that has the power to gather material from the blood and pour it out in the shape of fluid is known as a gland. Sometimes a whole organ does nothing else. Sometimes small glands are scattered through, or over, the surface of another organ. There are two kinds of glands in the skin of the mammal. The best known and most frequently thought of are those which pour out the perspiration. These have a double function. In the first place they assist in keeping the temperature of the body uniform. When we are too warm they pour out a watery fluid over the surface of the body. If the air is dry enough and our body not too closely protected by clothing, this perspiration passes off in the form of vapor. All evaporation requires heat, which in this case is extracted from the body. So soon as the temperature returns to its normal level the flow of perspiration ceases. The other function of the sweat glands is to take from the blood some of the waste matters of the body and pour them out upon the surface. This is done in order that the body may free itself from substances which, if they were to accumulate, would have a poisonous effect upon its action. It is this function of the sweat glands which makes it necessary for us to bathe the surface of our bodies with water. Dirt, in the ordinary sense of the word, is not harmful to a sound skin. Our reason for bathing is really to remove the wastes which we ourselves have poured upon the surface of the skin. These, if allowed to remain, soon decompose, like all nitrogenous substances, and become very offensive. They may then be reabsorbed into the skin and nature's effort to throw them off has been in vain. These glands, since they contain waste matter, could not possibly yield food for the young. They would poison and not nourish. Hence, whatever the breasts may be, they are not altered sweat glands.

There is another set of organs in the mammalian skin. At the base of each hair lies an oil gland. The function of these is to pour out a substance which spreads along each hair and over the surface of the body. The outside of the skin is always dead, and would easily crack were it not for the constant secretion of this oil. In winter, when the blood circulates less freely and these glands consequently pour out less oil, the supply frequently runs short. If what little is poured out is too frequently removed by washing, the skin becomes brittle, and, on bending a joint, the epidermis cracks. The gloss of the hair is due to the oil thus poured out. This oil becomes one ingredient in the milk produced by the transformed gland. But there is another important constituent. When one does unaccustomed manual work the ordinary result is the formation of a blister. The epidermis, or scarfskin, becomes detached from the dermis, or true skin, and the space between the two rapidly fills with the fluid portion of the blood, known as lymph. The fact that no blood vessels have been broken in this detachment results in there being no red corpuscles in this fluid. Wherever a cavity forms in the body lymph is liable to enter it.

The milk glands of the mammals are modified oil glands. The fluid which they now pour out is no longer exactly the old oil with the addition of the lymph. Undoubtedly in the past the first milk was more like this simple mixture. There seems no doubt that the breasts of to-day are the enlarged and modified oil glands of earlier mammals. In one of the most primitive of our mammals the young simply lick certain bare spots on the surface of the mother's abdomen. As higher forms arise there develops a smaller or larger mound with a distinct projection, about which the lips of the offspring can easily fasten. Lamarck would have said that the suction of the infant had produced such a mound, and that this had been transmitted to later offspring until it had grown to be the highly developed organ we now find, for instance, in the cow. Since, however, we have come to disbelieve in the transmission of acquired characters, this explanation will no longer serve. We must content ourselves with saying that, by whatever accident the nipple arose, the success of it when present determined its selection by nature and its consequent persistence. With increase in its function has come increase in the size of the glands. Lower animals which, like the hog, produce a large number of offspring, possess a large number also of these glands. With the diminishing number of young and greater care of them as we rise in the scale has come also a diminishing number of breasts in the female. Whether those on the front of the body should persist, or those on the rear, depends upon other factors in the life of the animal. Hoofed animals, perhaps because their best weapon is the hoof and they can there best protect their young, have retained them in the rear of the body. In the group of animals known as the primates, including monkeys, apes, and man, the habit of holding the young in the arms for protection has determined the persistence of the breasts upon the chest rather than the abdomen.

It is interesting to notice that the habit of the elephant of protecting its young by means of its tusks has also resulted in a similar position of the milk glands.

That the primates had once a larger number of offspring is confirmed by double evidence. Even to-day the number of children at a birth is often two, sometimes three, rarely four. The day before this was written came the report of a case of five children at a birth, all of whom seemed sound and all of whom lived. Still more direct evidence is found in the fact that occasionally in the human female there are two pairs of breasts, and very rarely three pairs. These are then disposed in a double line down the front of the body.

The new plan of caring for the young is one of the priceless heritages of the higher animals. As we rise in the grade of life the number of the young produced at one time steadily diminishes, while the care spent upon them increases. The shad may lay four hundred thousand eggs and trust them entirely to their fate. The sunfish will lay a thousand, by no means all of which can be fertilized, but it guards them somewhat after deposition. The toad lays several hundred, stores them with a considerable amount of nourishment, and protects them by a bitter deposit of mucous. The turtle has reduced the number of eggs to perhaps a score. Each of these is supplied with abundant nourishment, so that the young may develop to considerable size and activity before emerging from the egg. This material is enclosed in a firm protective shell and hidden away from sight by being buried in the ground. In the mammals comparatively few eggs are produced at one time. These are fertilized within the body of the parent, are attached to the parent, and absorb her blood. No shell is needed because nothing will kill the developing offspring that is not likely to injure the parent. Not only do the young feed upon the blood of the mother up to the time of birth, but they are practically dependent upon this same blood after birth. Though they do not take it directly from the veins, the milk is, none the less, the transformed blood of the mother. This assures the young of food as well as of protection. Best of all, the young are provided with the companionship of the mother. Now for the first time animals learn by example. Heretofore they have been born with a nearly undeviating instinct; now intelligence begins to arise. They can imitate their mother. Heretofore no acquired characters affected the young. In the mammals, although the young cannot inherit the acquired habits of the parents, they can get them by imitation, which serves nearly as well.

There is, however, a more wonderful advantage that comes from the close attachment between mother and offspring. This intimate relationship brings about an affection of the mother for her young heretofore unknown in the animal world. It is somewhat paralleled among birds, but here the care of the nestling is less intimate, far less maternal, than the care of the mammal for her young. As the number of the young grows less and the care taken of them increases, the intensity of the affection also increases. By the time we get as high as the dog or the cat this fondness becomes a fierce, self-sacrificing love. When we come to man, with his high intellectual powers, with his deeper moral sense, we find a wonderful change. This love of the mother for her child has grown into the finest emotion possible to the human heart. It no longer is confined to the dependent life of the child, but follows the offspring through its entire life, guiding, guarding, shaping its destiny, handing on to the child the treasured wisdom of the race. Influenced by the example of the mother, the father comes to have a love for his children. It is not so strong as that of the mother, nor so utterly unselfish, but it is still a noble and exquisite love. Developing in a different direction, the love of the mother for her children grows as civilization advances, and spreads over the father of those children as well. Again reflecting her love, the man finds himself filled with a new feeling for the woman. It is never as unselfish, as free from desire, as is her love, but it completely transforms his relation to her. What has been with him simply desire is ennobled and enriched until it becomes the finest passion of his life, absolutely transforming him, in relation to her, from a selfish brute into a tender and life-long companion. So utterly does the love thus engendered transfigure human life that when we seek to express the divine nature in human terms, and these are the only terms we know how to use, the richest revelation that has come to us is the conception taught by the Master that "God is Love" and that "as a father pitieth his children, so the Lord loveth them that fear him."

CHAPTER VIII

THE STORY OF THE HORSE

Ever since men have been familiar with the idea of evolution there has been a temptation on the part of the zoologist to draw up pedigrees expressing the relationship between the various groups of the animal kingdom. The impulse is natural, and, if the resulting tables are not accepted with too much confidence, the result is not undesirable. The truth of the matter is that all of these pedigrees are more or less hypothetical. They simply show what connection seems most likely. In all of them are spaces filled with doubtful names. Each addition to our acquaintance with the past history of animals necessitates revision of our tables. The student of fossils, trying to rebuild in imagination the world of the past, finds himself often strangely unable to link these animals together. The result is that the more we know of fossils, the more distrustful we become of the easy connections we have been making between groups. Accordingly we are more than commonly pleased when we find the clear indication of a genuine pedigree, actually illustrated by real examples, following each other in time through the geological history. A few of these lines are gradually becoming plain, and none of them is clearer than the pedigree of our familiar and much loved horse. The example is a particularly interesting one, not only because of our affection for the animal, but because the horse originated in all likelihood in North America on the land occupied to-day by our Western plains. As though he loved the country of his ancestors, he returned after having circled the globe, and once more went wild in the home of his forefathers. The problem was first worked out in Europe and later elaborated in this country. Now the history gets its finest expression in the American Museum of Natural History in New York City. The collection of fossil horses in that institution surpasses in completeness and in excellence of mounting and of sympathetic restoration any similar collection representing the ancestry of any other animal in the world.

In the table of Geological Times, given in chapter six, the era of recent life known as the Cenozoic is seen to occupy something like five million years. This figure, as was previously suggested, is very uncertain, and may be three or may be six, but is safely represented in millions. Through most of this time stretches what is known as the Age of Mammals, the Tertiary Age. Its close, occupying only the last few hundred thousand years, is known as the Age of Man, the Quaternary. Through perhaps three or four millions of these years stretches the known pedigree of the horse.

When we go back to the early Tertiary we find a forest, with trees that shed their leaves, interspersed with glades, in which already the grasses were beginning to be developed. This state of affairs had existed but for a comparatively short time, geologically speaking. It had come only in the latter part of the preceding era. Lake and swamp, meadow and forest intermingled to make a rich and varied scene. Slowly the land toward the western side of North America lifted itself into plateau and mountain range. Slowly the westerly winds began to be cut off by the barriers thus raised across their path. As they swept over the plateau and down into the eastern plain their moisture came to be diminished. Gradually a very different state of affairs set in. The ground became harder, the forest became sparser, the plants became higher and firmer, the grasses tougher and more wiry, and, by the time the Quaternary arrived, a condition probably even drier than that of to-day existed over our western highlands. Throughout this long change, spread over millions of years, a creature which has become our horse steadily persisted and steadily advanced. Side lines developed which finally disappeared, but the main line kept on, and when the Quaternary came the horse arrived with it. Many of the skeletons in this series were known before it was realized what they were. As time went on and intermediate forms were found, it became possible to recognize these as ancestors of the horse and to assign them their proper position in the family tree.

The earliest of the forerunners of the horse with which we are acquainted would certainly not be recognized as such by any but the most careful student of animals, if we could see him to-day. He stood not higher than a fox-terrier dog, though his shape was very different. But he would probably be more likely to be classed with the dog than with the horse by the hasty observer, for he walked with four toes of each foot upon the ground as the dog does to-day. Like the dog, he had hanging at the inner side of his front foot a little useless toe. He was long in body, comparatively short of leg, a little long of head and neck, and distinctly long of tail. His grinding teeth had points on them not unlike a pig's, and possessed no apparent resemblance to the wonderful curved and ridged surfaces seen on the teeth of the modern horse. What his skin and hair were like can only be conjectured. In the restoration which Mr. Knight has made, at the suggestion of Professor Osborne, an interesting inference has been drawn. That he was a creature of the forest is suggested by his spreading toes, which would keep him from sinking in the soft soil. It is consequently surmised that he was dappled with spots which allowed him to rest unnoticed on the sun-flecked floor of the forest. Mane he had none, and his tail was probably tufted slightly at the end with hairs, which were increasingly short as they approached the top. He had no forelock, and the hair along the ridge of his neck was a little longer than the rest, and stood erect. Browsing about on the soft and tender herbage of his woodland home, his teeth had as yet no tendency to become specialized. The molars had mounds upon them, developing, perhaps, more into the shape of the points of the hog's, but even still quite generalized teeth. His main enemies, from whom, perhaps, he could with little difficulty escape, were creatures related to the hyenas of to-day. Perhaps, like their modern representatives, they preferred eating their flesh tainted to exerting themselves enough to capture and kill their prey. By the time we advance a little further into the Tertiary, though still in its early portion, a remarkable change has already come about. The fifth toe, which in the earliest horse hung upon the side of the front foot, has completely disappeared. The change in the hind foot has gone still further. The hind leg in many animals evolves more rapidly than the front. The heavy work of running is always done by the hind feet, while the front feet serve rather as a prop to keep the animal from falling than as the actual means of locomotion. Hence the hind feet and the muscles of the hind quarters are almost always heavier than the front. Possibly on the front foot the little fifth toe was less of an obstruction, and persisted after the early horse had lost the corresponding toe on his hind foot. This process has gone on still further in this second stage, and the hind foot has but three toes, while the front still has four. This is not the only advance. Already the middle toe of the original set of five is becoming emphasized. The weight is thrown more forcibly upon it, as with the human foot it is upon the inner or big toe. The middle toe is growing larger and larger, and the nail upon it is spreading around it and is growing firmer. The creature, too, is standing more nearly upon his toes; his legs are getting longer; he stands higher from the ground, and now has come to be the size of a hound.

We can only surmise why this creature should have undergone such a change, but the presence of flesh-eating animals having the size of a fox, and presumably of the fox's swiftness, probably tells the story. The little bands of early horses, pursued by their carnivorous foes, were slowly modified into swifter creatures. It is not so much that running made them fast, as that the slow ones were continually being caught. If this process of constant elimination of the slow members of any herd is kept up long enough, the group will necessarily develop speed. As time goes on, of these early horses those which happened to have longer legs and stood higher upon their toes won in the race, and handed on their qualities to their long-legged descendants. As the animal rose upon his toes, the inner toe, corresponding to our thumb, was first raised off the ground and rendered useless, while a similar change came over the corresponding toe on the hind foot. The hard work of running being done on the latter, this superfluous toe was more detrimental there than on the front foot, and disappeared, consequently, more rapidly. In time, however, it also disappeared from the front foot. Gradually the further elevation of the foot lifted the toe, which corresponds to our little finger, off the ground, and this now disappears also.

With increasing toughness of the grasses, as the climate becomes drier and the region more elevated, the teeth of the horse are given harder work. The points begin to spread into ridges and to unite with each other in such way as to form the crescents, which are later to be so characteristic of the teeth of the modern horse.

By the middle of the Tertiary this ancestral horse has risen in height until he is taller and heavier than a setter dog. Three toes are found on each front foot. The middle toe is getting constantly more developed, though the smaller toes are evidently still of use. The ridges of the teeth are quite crescentic now on the outer side, and becoming better adapted to the evidently firmer food which the creature is obliged to eat.

As we come toward the end of the Tertiary, the development which had been all pointing in one direction has advanced very much further. The creature now would be undoubtedly recognized by anyone as a horse. The legs are longer and straighter; the middle toe has become the only useful toe, though on each foot a smaller toe, slender and probably useless, still hangs on either side. Two similar useless toes to-day hang at the back of the foot of the cow, which is now walking upon her two toes, which give her the appearance of carrying a cloven hoof. That is to say, the first toe on the foot of the cow has disappeared, the second and fifth hang useless and much diminished at the back of the foot, while the third and fourth are both well developed and serviceable in walking.

The late Tertiary horse has grown to be the size of a burro of to-day, though probably it was a little more slender. The teeth are quite horselike, both in shape of the crescentic ridges on their surface, in the length of the teeth in the jaw bone, and in the fact that the crinkled edges of enamel on the upper surface are protected on either side by dentine or by cement. These surfaces, being softer than the enamel, wore away somewhat more rapidly and allowed the sharp edges of enamel to stand up in ridges. This plan increases the grinding power of the teeth.

With the oncoming of the Era of Man the horse reaches his modern splendid development. During the early Quaternary the horse was perhaps in some of his representatives a larger creature than he is to-day. Each foot now has but a single toe. The nail has spread around firmly and heavily, until it has become a splendidly developed hoof, permitting the animal to travel with speed over firm and often stony ground. The side toes have disappeared completely from the outside of the horse's leg, although upon removing the skin it is easy to find the long splints, which are the remnants of toes, which have not yet quite disappeared. His heel has been lifted in the air until it is eighteen inches off the ground, and he is standing like an expert dancer upon the tip of his toe. The body of the horse thus being lifted far off the ground, a new development becomes necessary. All through the growth of the creature the neck and head have been obliged to lengthen correspondingly. Every animal must be able to bring its head down to the level of its feet in order that it may drink. Various animals use different methods to accomplish this result. The giraffe, with his enormously long legs, has a correspondingly long neck, which lowers his mouth to the ground. Even with this extended neck the giraffe's legs are so exceedingly long that he is obliged to spread his front feet when he wishes to reach the ground with his head. The elephant has pursued exactly the reverse plan. Using his tremendous head as a battering ram in fighting, and using his enormous tusks both in battle and in uprooting young trees, a lengthened neck is absolutely out of the question. Furthermore his front teeth have grown so prodigiously that they would interfere with his getting his mouth to water. Accordingly, his nose has lengthened its tip until it reaches the level of his feet, and this nose becomes to him the main organ of grasp and of touch. To drink, its end is inserted in the pool and water is drawn up the nostril. If the animal were to attempt to draw it all the way back into his throat, it would inevitably strangle him by getting into his windpipe. Accordingly, when the nose is well filled with water, the tip of it is inserted in his mouth, and the water discharged by a quick puff. The horse has taken a method intermediate between these. It had moderately lengthened both neck and head in order to get to the ground with its nipping teeth, and thus to gather the grasses which serve as its principal food.

The mammalian teeth, while of four kinds, really in most animals serve but two purposes. The front teeth consist of the incisors and canines, and are used for biting. The hind teeth, consisting of premolars and molars, are used for grinding. In the horse, the jaw has lengthened between these two sets, carrying the biting teeth far forward of the molars. It is this gap in the row of the horse's teeth which makes it possible for us to insert the bit into his mouth.

Now comes a strange accident into the life of our American horse. Creatures of the same kin had been evolving in Europe and Africa, but the developments are more distinctly horselike, it would seem, in our own country. Then for some reason the horse disappeared completely from American soil. Doubtless two things happened. First of all, some of them migrated across a stretch of open country which then connected America with Asia in the neighborhood of Bering Strait. These creatures spread first over Asia and then over Africa and Europe, leaving their skeletons scattered over this enormous stretch of country. Asses and zebras are still found abundantly and widely scattered, but the wild horse of to-day is seen only in western Asia. What happened to those who remained in America we shall possibly never know. Some surmise that a fly not unlike the tsetse-fly of Africa killed them out. Perhaps the members of the cat family, which are steadily growing larger and fiercer, fed on their young if not upon the older ones, and exterminated them. Perhaps the Glacial period which followed was too cold for them. But, whatever may have been the cause these horses died out not only in North but also in South America, to which country they had spread.

The old world horse was the companion of man. The skeletons of those found with early man in the caves of Europe look as if the horse had been a creature to draw man's burdens and to serve him for food, rather than to bear him upon its back. Its roasted bones are often found about the old tribal fires. Upon the discovery of the new world the Spaniards brought with them to Mexico and to the Mississippi Valley the horses which carried them in their battles against the Indians. In the course of these frays many riders were killed and their horses roamed wild. Slowly they made their way to the western plains; gradually they became tougher and more wiry; their diminished hoofs learned to catch more carefully in the rocks of their mountain home; and the mustang and bronco of more recent years are the descendants of the little dawn horse, whose dainty skeleton is found in the rocks over which his later descendants, after a long stretch of perhaps four million years, are now running.

CHAPTER IX

EVOLUTIONARY THEORIES SINCE DARWIN

In considering the value of Charles Darwin's work and its permanent effect upon the thought of mankind, we must be careful to distinguish between two phases of his effort. It was his aim to prove two propositions: first, that there is such a process as evolution; second, that he had discovered the method by which evolution is accomplished. Before his time there was no general agreement as to the fact of evolution. People generally thought the idea absurd, as well as irreligious. All previous efforts on the part of advanced thinkers to persuade mankind of the truth of evolution had been nearly without effect. Among the early philosophers the whole idea was purely speculative. They made no attempt to prove it, and the conception was without influence upon the thinking of the ordinary man. This remains true until the time of Lamarck. This French genius succeeded in persuading not a few people of the validity of the idea of evolution. He probably could have convinced many more had it not been for the hostility of Cuvier. Accordingly,

Charles Darwin's "Origin of Species" fell upon a world entirely hostile to the idea, when it thought of it at all. Within fifty years of the publication of this wonderful book, probably the entire scientific world is agreed that evolution, in some form or other, is the undoubted solution of the mystery of creation. The materialist may think of it as a mechanical process relentlessly working itself out without design or purpose. The theist will accept it as the plan by which Eternal Power steadily works. The devout Christian or Jew will see in it God's method of creation. The idea of development has penetrated every science that has to do with animals or man. It is even beginning to influence such inorganic sciences as Physics and Chemistry. We now hear of the evolution of the elements, and the evolution of forces. The world has been persuaded that evolution is true, and this is primarily the result of the work of Charles Darwin. It is astonishing that so great a revolution should have come in so short a time.

The other phase of Darwin's work was his attempt to find the agent which is bringing about the actual transformation of animals and plants. As we have seen in the preceding chapters, it was his idea that natural selection was the efficient agent which constantly eliminated all unfit variations, leaving only the best to carry on the work of the world and to reproduce their own fit kind. Many biologists since his time have doubted whether unaided Natural Selection will account for the constant advance in organisms. This is the part of the work which is often seriously questioned.

Weissman and his co-workers have contended that this unaided principle will serve. Most biologists have asked for some more efficient cause, and assert that selection does not account for the appearance of variations, but only for their preservation, and that any valid theory of evolution must show how variations originate. It is chiefly in this respect that Darwin's work has failed to satisfy many later biologists. When we hear a scientist speak of Darwinism as being dead, this is what he means. He does not think evolution false, but believes that Natural Selection is not sufficient to account for evolution. There are three main difficulties involved in Darwin's theory. The chief defect lies in the fact that selection cannot originate varieties. In all his earlier works Darwin simply accepted variations as he found them. He was content to say that all species varied constantly, and in every direction. He gave no theory to account for variation. Whenever he took measurements of the dimensions of any large series of objects of the same kind he found these measurements to vary, apparently, in all directions. Upon the facts of these variations, and without accounting for them, he built his own theory of evolution. He realized his weakness, and acknowledged it in his book. He probably did not anticipate how insistently later biologists would demand an explanation that would account for this variation. In his later work, responding to this criticism, Darwin originated a theory which he called Pangenesis. He believed that when an adult animal had responded to his environment and acquired a new character he could transmit this character to his offspring. At that time no one doubted this fact. The whole theory of Lamarck was based on the assumption that this could be done. Darwin suggested that every organ of the body threw off minute particles, which he called pangenes. These little bodies, carried by the blood, were taken up by the egg cells or sperm cells, and the latter cells determined the future development. Consequently, the character of the new individual was determined by the parental pangenes. In this way the gain acquired by one generation could be passed on to the next. This theory was purely speculative. He never pretended that there was the faintest corroborating evidence visible to the microscope in the organ, in the blood, or in the germ cell. It was not an accounting for what is, but for what it seemed possible to him might be.

This theory of Pangenesis, in the shape in which Darwin promulgated it, has dropped out of consideration almost entirely. DeVries of recent years has revised it, but with distinct modifications, and most biologists pay no attention to it.

There is a school of biologists, headed by Weissman, who have come to be known as Neo-Darwinians. These men have insisted that Natural Selection, if properly understood and developed, is quite sufficient to account for the fact of evolution, including the appearance of variations. Weissman himself is a microscopist of more than common skill. He is thoroughly accomplished in the most modern methods of killing, fixing, staining, and mounting. This worker's acquaintance with the intimate structure of the cell is probably as great as that of any other man in the world. Weissman asserts that he has seen inside the nucleus all the machinery necessary to explain how the father hands over his qualities to his children. He insists, equally strongly, that this process is such that no father can hand to his child any qualities which he himself did not have at least in potentiality at his birth. Everything the individual acquires during his lifetime is his own possession, which he may use and develop to the utmost extent, but it dies with him. His children, born after he possesses it, can no more inherit it than those born before. Weissman expressed this in his famous statement that "There is no inheritance of acquired characters." The biological world has had no shock equal to this since Darwin's time, and there are few other questions to which scientists to-day return with such constant vigor.

If what Weissman says is true, that no variation or development which comes to an animal during his lifetime can be transferred into his own germ cells and handed on to his children, then it becomes evident that we must find some cause of variation that acts within the germ cells. This is the difficulty which Weissman meets. He says that there are small particles in the nucleus of each cell; that these particles which he calls determinants decide the form and the course of development of that cell; that when that cell divides to produce another cell it gives to this other cell one-half of each determinant. As a result the second cell grows to be like the first. This tells us why offspring are like their parents. There is nothing in the theory thus far to show us why offspring are not exactly like their parents. In other words, there is no accounting, thus far in the theory, for variation. When the biologist studies carefully the history of an egg while it is being formed, he sees that at one stage in its development it throws away not one-half of each determinant, but one-half of the determinants. When an egg does this, it deliberately casts aside one-half of the possibilities of its own development. This throwing away is quite as effective for all its descendants. Any ancestral quality now lost is lost from the line forever. In the formation of the sperm cell set free by the male a similar throwing away of one-half the characters has taken place. The egg cell and the sperm cell fuse together. There are as many possibilities now as there were in either parent, but not all the potentialities of both parents. Half the possibilities of each have been thrown away, and hence cannot appear in the offspring. By this constant process we get, in every generation, new combinations of qualities. This is the main cause, says Weissman, for variations.

There is, however, another possible cause. Each cell has enough determinants in it for many individuals, and it seems to be more or less a matter of accident which qualities shall come out. It has been suggested that as an egg lies within the gland, a blood vessel may bring blood to it in such way that a determinant, lying in a certain position in the egg, may get the richest supply of blood, and hence develop at the expense of the less nourished determinant. By these two methods variation comes into an animal's life, if Weissman and his school are to be believed.

This is a serious blow, if true, to many theories of evolution. The great mass of evolutionists still feel that somehow there is an influence by which the environment produces variation. How the influences of the surrounding world can get down into the body of the parent and affect the egg is unknown. This is freely confessed by every biologist. All are agreed that Weissman's work has made us cautious, and prevented our lightly accepting a belief in the influence of the environment. Yet it is felt by many that slowly and gradually, in the long run, the germ is affected in the same manner as is the body of the parent. In other words, even those who are not followers of Weissman, have accepted the idea that there is little inheritance of acquired characters. Yet they return to the belief that somehow, in some way as yet unexplainable, the main cause for variation in animals lies in the situation in which they live, and tends toward better adaptation to that situation.

Whether men with this conviction are merely reactionaries whose confidence is returning, or bold thinkers whose views will ultimately prevail, time alone can tell.

A second strong objection was brought against the theory of Natural Selection. Darwin declared that small variations in favorable directions are selected and become the starting point of new and better things. It is soon seen, however, that the effect of unaided Natural Selection would be but to mix new departures with the old forms, and soon swamp out any progressive tendency. Whenever a genius appeared, instead of finding a corresponding genius with which to pair, it mated with the average of its own species. Hence its offspring were nearer the average than it was, and their offspring still nearer. Thus whatever advantage the genius originally possessed gradually sank into the common level.

It was Moritz Wagner, a German naturalist, who first insisted that if favorable variations were to amount to anything these possessors must not only mate with others of their same kind, but must also be prevented from mating with the old average group. Accordingly, the belief arose that, under ordinary circumstances, variations returned to the common level. Wherever a varying group became separated by any barrier from mating with the rest of its species, and had only its own kind to pair with, a new species sprang up. This barrier might be a desert, or an impassable mountain range, an arm of the sea, or anything else that the animal could not, or would not, cross. Isolated in this way, the little group that had an advantage in a different direction could develop its tendencies, and a new species would be made of what had been previously only a geographical race. In this matter of geographical isolation Wagner is very strongly supported by the American zoologist, David Starr Jordan, who believes that no two closely related species of animals ever occupied the same geographical area. Both Wagner and Jordan are ardent admirers of Darwin and his theory of natural selection, but both believe that it is necessary to add the idea of isolation in order to make natural selection effective.

George John Romanes, a British naturalist, has added to Wagner's idea of isolation, the expanded conception that there may be isolations that are not geographical. For this phase, Romanes has coined the term physiological isolation. Something in the structure or habit of the animals with the new variation prevents them from mating with the older type. Occasionally it is a difference in the structure of the reproductive organs themselves. This, however, is not the only possible divergence. The mating season in one group may come earlier than that of the other, or may come during the day, while the main group is in the habit of mating at night. Anything which keeps some members of a species separate in their mating from the rest, will result in the course of a longer or shorter time, says Romanes, in the formation of a new species.

A third great objection was raised against Darwinism. The theory said that only useful variations were selected by nature. It was asserted by objectors that the earliest beginnings of any variation must be too slight to be useful, or as the term went, to have selective value.

It has been noticed by a number of naturalists that certain animals seem to carry the development of a peculiarity altogether too far. It is seen for instance that in the Irish Elk, which has for some time been extinct, the horns were so enormous as to be a source of danger rather than of assistance to their owner. It was said that the tendency to produce heavy horns had gained, as it were, a sort of momentum, and that this impulse had carried the development beyond a safe limit. The Irish Elk became extinct because his horns were too heavy. During the Mesozoic period the reptiles grew too large. They seemed to have carried size to a point at which it became a danger instead of a help. They completely passed out of existence, leaving behind them only very much smaller reptiles.

Eimer, of Germany, has based on facts like these his theory of Orthogenesis. He says that variations in animals are not indefinite and in every direction, but that they follow along clear and definite lines. These lines, in the case of the elk and of the Mesozoic reptiles, developed too far, but ordinarily the effect of such a tendency is distinctly beneficial to the animal. It particularly assists in carrying on for a time the variations which have not yet become useful to the animal. It has always been difficult on Darwinian principles to understand how the beginnings of the useful variations could be selected before they were strong enough to be of actual value to the animal. This tendency to variations in certain directions instead of at random would account for such early development. This theory of Orthogenesis has not figured very strongly in the history of the movement, but it recurs at intervals.

Both in America and France there is a constant tendency on the part of zoologists to return to the Lamarckian idea that it is the use of an organ that develops it, its disuse that makes it fade away. This is undoubtedly true of the individual, and although Weissman insists that it is useless to the species as a whole, many zoologists are slow to relinquish entirely the idea that somehow these favorable developments become reproduced in the offspring.

Professor Cope, the American paleontologist, was a strong believer in the effect of activity, both upon the individual and upon his descendants. He believed that the insistent beating of the foot of an animal upon the hard soil of the drying Tertiary plateau, had influenced the production of a firmer nail, which spread around the entire end of the toe and made the hoof of the ungulate. He believed that the use of the teeth in grinding produced a stronger and better molar tooth, and that the offspring shared in this advantage. Since Weissmann's time, however, every Lamarckian feels it necessary to suggest some method by which the altered body of the parent can produce modifications in the germ plasms from which the young are to spring. One of our later biologists begins to talk of some effect comparable with wireless telegraphy or induced electricity. He believes that organs in the adult, not necessarily by direct action, but by action from a distance, may alter the germ. Of this, there is no proof at present. Others have suggested that just as the ductless glands pour into the blood chemical substances which materially affect the growth and development of other portions of the body, so similar enzymes, or other chemical substances, may be sent into the blood, which subsequently bathes the germ cells of the coming generation and produces the change. But of this, again, there is no proof. We may believe that acquired characters are transmitted, but we certainly do not have a very clear idea as to how it can be done.

One of the strongest objections to Darwin's idea of evolution by natural selection of small and favorable variations, is that the process is too inconceivably slow to account for the enormous progress which has been made. The answer has always been that our observation ran back so short a time that we really have no clear idea of how rapid evolution may have been. Again, it has been answered that transitional geological periods, in which there is much change in the physical geography of a country, will produce more rapid evolution than we at present are experiencing.

Hugo DeVries, of Amsterdam, believes he has found the answer to this difficulty. Outside of his botanical garden an American species of Evening Primrose had run wild. In looking over a number of these plants he found, every here and there, certain peculiar members of the species. They differed noticeably to the practiced eye from the rest of the group. When they were planted and crossed with each other, and the resulting seeds were again planted, the peculiarity remained constant in all the members of the collection. Here then we have a true variation, not large in amount, but at the same time quite definite, and which from the first remains true. Here are the beginnings, says DeVries, of new species. They are true from the first; they can live among other members of the species and still come true; they do not need isolation, at least in Wagner's geographical sense. These forms DeVries calls mutations. It is his thought that a species may run along uniformly for a long time when, from some cause which he has not determined as yet, instability comes into the species and it varies in quite a number of directions. Each of these variations may be the starting point of a new species. DeVries believes that he has at least half a dozen mutants of his new Evening Primrose.

This theory of Mutation has been eagerly seized upon by many botanists. The zoologists have not accepted it quite so enthusiastically. If this is the chief method by which species transform, it seems strange that we do not find more mutations than we do. Perhaps we do not look carefully enough; perhaps we shall find them a little later. Just at present it seems premature to believe that all evolution is by mutation, although quite possibly some of it is. The main apparent advantage of mutation is that it hastens the time in which a new species may arise.

There are certain difficulties which run back into the problem, and which must first be reasonably solved before a clear understanding of the idea of evolution is possible. The first of these is as to the nature of life. What is life? The reply of the biologist will probably be that so far as its material side is concerned, it must be answered in terms of physics and chemistry. As to any side not material, if it have any such side, science says that the chemist can have nothing to say. The chemist may have an opinion of his own based on some other ground than his chemistry, but so far as he is a chemist, he has no opinion. The chemical side of life is being very carefully and very fully investigated. We are certainly being brought nearer to the borders of the living substance. We are rapidly gaining fuller knowledge of the physical and chemical processes which constitute life, or with which life is always associated. If we gain this knowledge we shall be in better position to solve many of our other problems. Even then there is a problem which preceded and which will possibly always defy solution. How did life originate? Has it developed out of chemical and physical activities which we know as heat, light or electricity? If so, what were the conditions under which it developed? If we understand the nature of life, and the conditions under which it developed, we may be able to produce it at will.

A few scientists may hope dimly that this will be attained. I suspect a great majority believe it to be impossible, and that the question as to whether life evolved upon this planet, or this planet became infected with life through meteoric dust from some other center, will forever remain an unsolved problem.

CHAPTER X

THE FUTURE EVOLUTION OF MAN

The disturbance of mind created by the publication of Charles Darwin's "Origin of Species" would have amounted to nothing if the theory had been applied to the lower animals alone. Few people would have disputed that a cow and a buffalo had descended from the same ancestor, or that monkeys and apes were of a common blood. The whole theory would have been looked upon by those outside the biological world as entirely an academic question, in which they had little concern, and less interest. But within this century the scientist has so persuaded the world of the unity underlying the activities of the universe, that so soon as a principle is established men begin to run it out to the very end. Everyone knows perfectly well that if it could be proved that the dog and the horse had a common ancestor, still more if it could be made apparent that the dog and the frog and fish had sprung from the same stock, then there could be no question of what would be the final application of the theory. Man himself could be no exception to the law. So the battle dropped at once upon this most interesting point, and around this center the contest has waged.

What is the origin of man? Who are his ancestors? As soon as we ask the question there is no doubt whatever as to the answer, if we accept the principle of evolution. Our only means of judging relationship between animals is by the similarity of their structure. As soon as we come to examine the other creatures even in the most cursory fashion, there is only one group which in any close degree resembles the human species. Our nearest relatives among living animals must undoubtedly be the apes. Some little distance farther away stand the monkeys, and, structurally speaking, there is more difference between a monkey and an ape than there is between an ape and man. The gap between man and his relatives of this group, known as the primates, is a mental, not a physical one. While his brain and his mind have developed far beyond theirs, the rest of his body is comparatively close to that of an ape.

Probably no one can face the possibility of his being descended from creatures not unlike the ape, without feeling a stirring sense of repugnance. The least aristocratic of us hesitates to name in the line of his ancestry creatures so unlike himself as the members of this group. It seems to us impossible that we should have descended from creatures as lowly as they. If evolution is true, these are among our near ancestors. Back of the group of primates lies a far less developed set of insectivorous animals, behind them the reptiles, behind them the fishes. When we get back this far we are less certain but most probably the worms take up the story. So our ancestry runs back to the very beginning, when it originated in the one-celled animals which are also the ancestors of all the rest of the animal world. If we are inclined to deny our ancestors in the trees, what shall we say of our forefathers in the seas?

The question of course is not to be decided by our likes or our dislikes. If the evolution of man is true it will not make it less true because the process is not to our liking. It is our part, if this be the truth, to accept it as we do any other truth. Surely those of us who are moral of thought are not willing to disbelieve a truth because it is unpleasant.

The newness of the idea is the chief reason for our dislike of it. This lowliness of origin should not be distasteful to us. Nothing about Abraham Lincoln seems to us more wonderful than that a man who towered head and shoulders above his generation, indeed above most generations of men, in his fineness of life, in his nobility of purpose, in the integrity of his aims, should have been of exceedingly humble extraction. It only adds to the glory of his later achievements that he should have lived in a cabin, have spent his young manhood splitting rails and running a flat-boat, and have gained his education almost unaided from a few books and much meditation in front of a log fire.

That the greatest military General on the Union side of the Civil war should have been the son of a country tanner, and as a boy, not over-shrewd in the matter of bargains, adds to the glory of his later life. The simplicity of his childhood gives new luster to the power with which he led the forces of a nation to victory, and then went to a battle no less noble in his long fight for honor while suffering from disease and approaching death. Why then should we feel that such beginnings in the lower world are too humble for man? Why do we think his present superiority diminished by his lowly origin? Why can we not see that precisely the reverse is true? The more humble the level from which he sprang the more gloriously creditable is his present position. Instead of being ashamed of having risen from the brute, it should be the glory of man that he has so sprung. His chief superiority lies in the fact that while they have remained where they are, he has so completely outdistanced them as to have placed a gap between himself and them that seems almost impassable. Furthermore, if man with his present glory of intellect and of moral impulse, has sprung from a creature whose superiority to the ape lay chiefly in its potentialities, then it does not yet appear what he shall be. We can judge the future only by the past. Through the long ages the development has been very slow. Through the last hundred thousand years the development of man has been wonderfully rapid, compared with what went before, though it seems slow enough when we look at it from the standpoint of our historical and traditional reports. But with this added impulse, this rapid improvement that has come with the development of mind instead of muscle, of tooth and of claw, we have every promise of an evolution that shall far surpass anything that has yet come. To-day our leaders are way beyond the average of the mass. Who shall doubt that in a not too distant to-morrow, the masses shall be where the leaders of to-day now are. We shall not then have reached a dead level of superiority. Our leaders will have moved on as rapidly as have the masses, and will be as far ahead of them then as they are now. It shall be their work to apprehend new virtues, and to work them out in their lives. The masses, seeing the beauty of the lives of the leaders, recognizing in those lives the revelation of the divine power which they have apprehended, will hunger to learn of them and to lead lives like theirs. To this process who shall set an end? The advance is slow, as in all evolution; but anyone who wishes to do so may easily detect the direction of the current.

The evolution of man's physical frame probably has nearly ceased. Gradually organs that are useless to him are passing away. Slowly his hands are becoming more delicate and refined and skilled. But his evolution has begun to work itself out on entirely other lines. We sometimes hear that the men of the past were the full equivalent of the men of to-day. Scholars like to tell us that the population of Athens was finer in quality than any population that has existed since. We must remember that group after group of men may be expected to specialize intellectually and fail to develop morally and physically. Under these conditions this little branch of the human race runs through its forced flowering and comes to an end. With the study of history and the earnest investigation of these lives of the past, new possibilities arise within the human family. The next race that flowers may take longer to decay because it understands better the weaknesses that carried away the preceding civilization. In time there will arise a civilization that understands the past. A whole people will some time realize that intellectual development alone will not save it, or Athens would have lasted; that moral development alone will not suffice, or Judea had been permanent; that physical development will not serve, or Sparta would stand to-day. Some day there will arise a nation that will see to it that every intellectual advance is accompanied by an equivalent moral and physical advance. When this time comes we shall have a race which can survive. Are we to be that race? The sins of man are generally the dregs of his brute ancestry. Bestiality of life was once common enough to attract no attention. Kings and nobles were not supposed to be clean so long as they confined their bestial relations to those below them in rank. Gradually men are becoming ashamed of uncleanness in life. Some day there will be no difference so far as purity of life is concerned, between the two who present themselves at the altar asking the blessing of God on their union.

If anyone doubts that English speaking people are becoming cleaner of life he needs only to consult the literature of the past. No one dreams of finding fault with Chaucer because his stories related in the company of men and women often would not bear such telling to-day. Shakespeare, with all his wonderful genius, needs expurgating if one would read him aloud comfortably to a mixed audience. And these are the shining stars. When we drop below them, the literature of their time becomes nearly impossible to read. Fielding and Smollett and Stern helped to build up the English novel, but the stories they tell speak of the grossness of their time in language that is unmistakable. We are by no means clean to-day. A fair proportion of our novels leave much to be desired. The stage is the scene of much we could wish to see cleaner. Above all this grossness there towers a sweetness and beauty of thought, and an earnestness of purpose, a sincerity of effort, which makes the present time fuller of moral purpose, fuller of the desire to be clean and to help others to be clean, than graced any previous period in the history of either England or America.

Under the change from country to city life man has suffered. Here too evolution is necessary. City life tells hard on the second generation and nearly destroys the third; but we have come to understand the difficulty and are fast remedying it. It is more than possible that the next generation will see such changes in the life of the worker in the great center, as shall effectively stop the physical deterioration that has come to the city dweller. God grant that modern civilization has had teaching enough and learned its lesson well enough. God grant further that we may give over slaughtering our most ambitious and vigorous young men in battle to settle questions which battle can never settle. God grant that we have come to a turning of the ways where the life of men, women and children, no matter how humble their station, shall stand higher in value than the profits of any commercial venture. God grant that we will soon be firm enough to declare that a business which can only live by sacrificing the health and strength of the workers must be counted an unprofitable business, and be allowed to cease. God grant finally that the American people may learn from the past to guard against a like fate in the future; that here may be the people whose strength, intelligence and uprightness shall lead the world; not for the sake of exceeding the world, but with the high mission of setting to the world an example of what can come to a vigorous, free and God-fearing people.

In the early history of the evolution of man the struggle almost always concerns the individual. Gradually the family comes to be the fuller unit. Only that is success which leads to the success of this higher group. After a time the family broadens to the tribe, and then the tribe to the nation. The evolution of social institutions is at present going on at an enormously rapid rate. Throughout the civilized world democracy is coming to its own. Even where the form of monarchy still prevails, the subjects of the monarch are having more and more rights. The people of England are surely as free as are the people of the United States. Increasingly all forms of government will secure for all their subjects, no matter what their station in life, a fair share of the general prosperity. In this field, human evolution is perhaps more rapid than in any other.

Any individual human being is a network of traits and peculiarities. He has all the ordinary attributes of humanity, but to the whole complex he gives an individual peculiarity which is totally his own. Where did he get his qualities? In the earlier times the fairies were supposed to have blessed him or cursed him in his cradle. A later age saw in the stars the rulers of man's destiny. He was jovial, or saturnine, or martial, depending on the planet which was in the ascendant at the time of his birth. Now we know "it is not in our stars but in ourselves that we are underlings." Everything a man is comes to him from within or from without; from nature or from nurture; from his heredity or from his environment. From our ancestors we get all the possibilities of our lives. To a certain extent we are slaves to our heredity, but not by any means to any such extent as to make us hopeless, unless our heredity is miserably bad. To the great mass of us come larger potentialities than we ever develop, and such possibilities of degradation as, fortunately, few of us ever reach. Within an enormously wide range, man is the architect of his own fortune. Only such traits develop as find a stimulus in the environment. Accordingly, a very large proportion of the development a man may achieve depends upon the circumstances under which he is placed, or, what is far more to the point, in which he may place himself. Man is not the blind sport of a relentless destiny. It is his to choose his environment; it is his to modify his environment when he cannot leave it. To an extent which no other animal has ever approached, man is the arbiter of his own destiny. A hypothetical ass may stand helpless between two equidistant bales of hay, but no human being is ever so helpless a sport of his environment. As it is, he may drift or he may rove as he pleases. To one man the current may be stronger than to another. There may be now and then a child so feeble-minded as to be unable to decide the course of its own life. It will not be long before society will see to it that such a life leaves behind it no strain cursed with its fatal weakness. In this effort to advance, man has all the advantage that comes from concentrated social effort. No man may live to himself. To every man in our community who desires it, a helping hand will be stretched. Often a hand will be stretched to him and he will be steadied whether he will or not, until his own will reforms itself and gains the mastery.

Inasmuch as all that is in man comes from his environment or from his heredity, the only way in which the race of men can be advanced is by improving their environment or by bettering their heredity. The first of these is the province of the sociologist; the second that of the eugenist. The sociologist has for some time been giving his careful attention to the improvement of the environment. In every large city, a man must build for himself a house fit to live in, if he build it at all. Whether he erects it for himself or for another makes no difference. Society will no longer allow him to build a home which is a detriment to the one who lives in it. Not only must he make himself a decent home but he must keep it in decent condition. The community will not allow him to endanger his own health, or that of his neighbor, by an insufficient method of attending to his garbage, or by a lack of ordinary cleanliness. If he will not clean his premises himself, the law sees to it that they are cleaned for him. Already we are beginning to understand that no man has a right to employ another man or woman or child at wages which are not sufficient to maintain the one thus employed. The wages of many people are exceedingly meager, notably those of women and children. He can read but ill the signs of the times who does not foresee an early end to the exploiting of the labor of these helpless creatures. Humanity has determined firmly that these things must pass, that the young child must not labor long or hard, that a woman must not be taxed beyond her strength. Already in England there is a partially successful movement which will doubtless spread to this country to provide that a woman be granted a little time before and after the birth of her child during which she shall not be allowed to suffer because her power to earn a wage is temporarily gone. These things cannot fail in the long run to strengthen the people. They strengthen chiefly the present generation. The blight of the fact that acquired characters cannot be transmitted, meets us here. This improved environment can only slowly, if at all, improve the race, and every effort made in this direction must be repeated with each generation.

Under such circumstances is it to be wondered at that the eugenist is hoping to raise the strain? Any improvement he can bring about is not only valuable for the generation in which it comes but is carried on into the generations which follow. This is the hope that strengthens and sustains him in his effort. The science of eugenics is so new, so little is surely known concerning the transmission of human characters, that no one is able as yet wisely to say what course is to be pursued in improving the race. But the problem is so interesting and its outcome so overwhelmingly important that men will never cease striving to know, and may, before many years, begin wisely to guide us in our efforts to provide a finer stock.

Heretofore our efforts at improving the strain have been confined to cattle, chickens and plants. An almost unalterable repugnance rises as soon as we speak of improving the human strain. Visions, if not stories, start up at once, of experimental matings of human beings, and of all other unspeakable abominations which no decent man expects to happen or even wishes to attempt. If there is one thing in human society the value of which has been demonstrated through the unending ages, it is the monogamic marriage. All ideal workers must point to the life-long union of a strong, vigorous, clean-minded and clean-lived man with a similarly fine, strong, clean-minded and clean-lived woman. Such an ideal may be slow in its attainment, but he aims too low who aims to secure anything less than this. The long struggle out of bestiality into pure monogamy has been so slow, so gradual, so noble in its attainments, and is still so far from perfection, that it would be an inconceivably stupid blunder to let go a single point that has been gained. Whether divorce shall be allowed to remedy a mistake may be a matter of dispute, but at best it is a bad remedy for a mistake that should never have been made. No ideal society could ever consider divorce as any permanent portion of its activities. Children are not like cattle. It is not simply a question of their being brought into the world sound and strong. Their long infancy which in the biological as well as in the legal sense, lasts until they are grown up, should be spent in surroundings which can minister, by example and precept, to moral and intellectual development. Surely no such end can possibly be attained when man and woman mate lightly, to part quickly.

At first sight it would seem a wise thing to require health certificates for those who would be married. I doubt not the Chicago Bishop who declined to marry his parishioners except under such conditions, will exert a beneficial effect upon the country by the attention he thus attracts to the subject. It would be a bad day for the city if all the clergy and all the other authorities who are authorized to solemnize marriage should take this step. We have not yet arrived at such a stage of development that a marriage certificate is essential to mating, and a restriction of this sort would simply mean that there could be no legitimate union except of those in strong health. To the burden of ill health would be added the still worse handicap of an illegitimate parentage, with all its bitter train of scorn and shame. Accordingly, it must be possible before the law for those who are not thoroughly vigorous to marry. But, year by year, we may come nearer accomplishing a finer mating by the aims and purposes we foster in the growing generation. Marriages will never be worth while when they are not freely entered into by the contracting parties. Choice must be free and unrestricted if it is to last for life; but this does not mean that it must be unguarded. It would be bitter folly for parents to leave to their children, without attempt to influence or restrain, the making of their marriages. The mating of our children must be inspired, not directed.

There is one taint from which society has the right and the duty of freeing itself, so far as in its power lies. This is the taint of feeble-mindedness. Of all the calamities that can befall a human being, feeble-mindedness is, perhaps, the worst. From most misfortunes it is possible to recover; with most of the rest one may exist without detriment to the race. To be feeble-minded simply means to hark back to the level of our animal ancestors, without regaining their power to guide life. The animal is provided with a bundle of instincts which tell him what to do in all the ordinary emergencies of life. The human species, in its development, has lost a large portion of its instincts, and has gained, instead, the power of intelligent choice and the ability to learn by imitation. When these drop away, man without his instincts or his intelligence is more helpless than the brute. Students of sociology are making clear to us that a large portion of the criminality of the world, much of the looseness of life, and a large part of the alcoholic excesses are due to this taint of feeble-mindedness. Recent investigations have made it clear that one feeble-minded family in a community may, in the course of years, poison the life of an entire state. The Jukes family in New York, the Kallikak family in New Jersey, have shown the awful possibilities of descent from a single feeble-minded ancestor. Prisons, almshouses, and houses of shame owe their population in no small degree to this bitter curse. It will not be long before society will learn to protect itself against such poisoning of the human stock. Nothing is more clear to the investigator of this subject than that the one overwhelming cause for feeble-mindedness is feeble-mindedness in the parentage.

There is one type of mental weakling, known as the Mongolian idiot, which may arise right out of the heart of an apparently sound family. But the number of these is comparatively small. The number of feeble-minded, who are feeble-minded because of their heredity, is dishearteningly and astonishingly large. Every attempt to examine large numbers of school children shows a sickening proportion of those who are distinctly feeble. Every little community seems to have its boy or girl who is what is known as silly. Such people rarely live long lives without leaving behind them feeble-minded children, no small proportion of whom are likely to be illegitimate. Against this fouling of the stream at its source, society must protect itself. Legislators revolt at the somewhat inhuman but certainly safe method of surgically preventing the possibility of the feeble-minded becoming parents. It would be more creditable and just as effective if society would take upon itself the tremendously expensive task of caring for all its feeble-minded in institutions during their entire life. The cost would be large for a generation, but would rapidly diminish and eventually become small. It certainly would be the humane way. These people in good institutions are by no means unhappy. Within the limit of their capacities they can do many things. Wise management usually will secure from them labor enough of wholesome and simple kind nearly to pay for their own support. Nothing could be better for them than to till the soil, care for the cattle, tend the chickens, and, in this way, provide very largely the materials on which they are fed. How this problem shall work out, time only can decide. With it once worked out, there is no doubt that the level of humanity will be distinctly raised. No other one feature in the program of eugenics seems more absolutely hopeful than this.

In several of the states of the Union it has recently become the practice to remove the possibilities of parenthood from certain classes of criminals. The purpose of this is clear and benevolent. Society has a right to prevent the oncoming of new generations of foreordained criminals. Underlying the practice is the theory that the children of criminals are born criminals. It is far from likely that this is the case. Criminality may be due to a wide range of causes. If the criminal is one of those actual born degenerates whose whole mental and physical make-up is so defective that nothing but criminality can be expected of him, then we have a case in which it is clear that society may, and should, remove the possibility of having more generations of the same kind. Probably only a moderate proportion of the criminals in our jails and penitentiaries belong to this class. Doubtless a distinct majority are criminals more through environment than through heredity. Born of average ability, or more, these people have been criminals simply because they were reared among criminals, because their surroundings were such as to lead them away from habits of industry, while they must live. These people were not bolstered by society, or the church, into a life of self-respect and self-help. Under these circumstances they fell into evil ways. There is nothing defective in their mental or physical make-up, that need appear in their children. If these children are removed from contact with the criminal class they stand every chance of being as vigorous, as intelligent, as upright as the average of the community.

At the recent Eugenics Congress in London one of the speakers expressed a preference for the son of a husky burglar over the son of a tuberculous bishop. This is doubtless quite correct, but why should the bishop be tuberculous? The truth of the matter is, the reverse is more likely to be the case. Personally, I should prefer to be the offspring of a husky bishop. In dealing with criminals, then, with a view to cutting off their posterity, we must be careful to understand whether we are dealing with a hereditary or an acquired criminality. If there is a genuine hereditary criminal taint, society is right in freeing itself of it. If it is acquired criminality, then it is not transmissible, and the offspring, if placed in a good environment, are likely to be good citizens. All of which means that, until we are clearly sure of what constitutes a hereditary criminal trait, we should move very slowly in the matter of mutilating criminals.

What steps may the eugenist, with his present limited knowledge, clearly, hopefully and confidently take to improve the future of the human species? There is one avenue open to us in this matter in which we can hardly go wrong. Even our mistakes can work little harm, and every well-done piece of work in this field will be a blessing to the race. This step lies in inculcating in our boys and girls high ideals of parenthood. This is more effective than legal prohibition of certain forms of marriage which cannot prevent matings, and adds the curse of illegitimacy to the other handicaps of the children of such unions. The first step in this process has already been reasonably well accomplished. Both our boys and our girls are in love with health. A good husband and a good wife should be healthy and vigorous. This does not mean that we expect a boy or girl who is looking forward to marriage to sit down and ask himself deliberately about the health of the person with whom he would mate. We must fill our children with the love of outdoor life, with the love of exercise. This will foster in them an admiration for people who are vigorous of body and alert of mind. It ought to become practically impossible for a hearty and vigorous boy to fall in love with a helpless and an 鎚 ic girl. It should be equally impossible for a hale and active girl to admire a man who was her inferior in either vigor or alertness. The modern taste for outdoor life has largely brought this to pass among such of our people as have leisure enough to indulge in vigorous sport. Among the crowded dwellers in the closer sections of the city such life has been so nearly impossible that no ideal of vigorous manhood or of radiant womanhood has had a chance to grow up. With the oncoming of the parks and play-grounds, all of this, we may hope, will change. Health and vigor will be no less attainable and hence no less adorable in the city than in the country. Rich and poor alike will be attracted by rosy cheeks and an elastic gait.

Our aim, however, should not cease with a vigorous body. We must teach our young men and young women the glory of a well disciplined mind. This should seem quite as admirable to them as a vigorous body. To them, straight thought ought to be as lovable as a firm and supple body. In this matter our young people are less exacting. The ordinary conversation of people gathered together for social purposes is not particularly intellectual, and any attempt to make it so at present seems priggish. With a broader education, will come keener demand for intelligence. We may hope the time is not too far distant when a question of governmental policy, a new book or play, or a new discovery in science will stimulate as much conversational zest as now seems to be gotten from a pack of cards.

A third feature of the ideals which should be instilled into the minds of our children is the moral phase. There seems little doubt that this is on the way. We must not mistake an evident laxness of religious observance as being synonomous with moral looseness. The revelations which our recent periodicals have brought us concerning the habits of business men, of politicians, and of society, have left on many minds the impression that this is distinctly an age of decadence. Exactly the reverse is the truth. This is the age of intense sensitiveness to wrong. In almost no particular is it worse than any previous age in the history of our country. We openly discuss things which we left untouched a little while ago. We insistently demand that business practices to which nobody particularly objected a dozen years ago must now certainly cease. All of this has produced an erroneous impression that the times are out of joint. But the dust and dirt in the air is the unavoidable accompaniment of house cleaning. When doubtful practices simply have publicity many are awakened to the sense of their duty to society. Persons who, of themselves, might be willing to live low and godless lives, dare not do so in the face of society when our social ideals are finer. I believe there is the utmost hope that within two generations our young men and young women will scorn meannesses which we are accepting with entire complacency.

A close acquaintance with thousands of young men and young women running through an experience of twenty-five years has taught me to believe that our young people of to-day are altogether cleaner of mind, of tongue, and of life than were their parents. There is freer, franker discussion of many things that their parents would scarcely have dared mention, yet I feel sure the moral tone is distinctly higher. I look with entire hopefulness to an early season when the young man who asks a woman to share her life with him will be met with the entirely proper question, "Have you kept your life clean for this event?" I believe that unless the answer can be in the affirmative the young woman will not be able to have admiration enough for the young man to cover uncleanness in his life.

There is one temporary phase of present life which seems discouraging. The increase in the cost of living, and still more rapid increase in the standard of living is shifting too late in life the age at which our young people marry. The result is that one of two things is likely to happen; either a large number of people are likely not to marry at all, or the romantic time of life is passed before the event occurs which it is intended to bless. A young man and young woman who are in this time of life can deny themselves for each other, can struggle and plan together, can hope and trust together to an extent that can never be the case if marriage is delayed beyond the romantic years.

The best foundation possible for a life of happiness is vigor, ability and good character. For the lack of none of these can wealth properly atone.

There is an apparent tendency to waken to the situation. I hope it will come soon enough for our young men and young women to get past a desire for such establishments in life as their parents already have. With this difficulty removed, with our widespread education, with the constant diffusion of both information and ideals from our periodical press I have every hope that the evolution of a new, a finer, and more vigorous race, will come with a rapidity which nothing that the past has done would lead us to expect.

CHAPTER XI

SCIENCE AND THE BOOK

We of the twentieth century have an overwhelming desire to be up to the times. Nothing but the latest news on any subject will completely satisfy. We are more anxious for late information than for accurate information. We have an almost unconquerable feeling that if it is late it must be accurate. All of us are sensitive to being thought behind the times. We feel that no stigma can be more invidious in the intellectual world than the stigma of being out of date. This pervades the masses quite as strongly as it does the more cultured classes. Under these conditions everybody wants to know the latest theory that science has to offer concerning anything that can be brought within the range of their interests. As a result everybody would like to know about evolution, were it not for the fact that a great mass of people have been brought to believe that there is something inherently irreligious in the idea. Our people have a saving sense of the value of religion. Denominational control may set lightly upon them. Certain long revered doctrines may have little practical influence upon them. Yet inherently they all believe in religion, and most of them believe themselves to be religious, as indeed they really are.

It is a most wholesome tendency which leads us to esteem religion as the main interest in life. We must feel a sense of shame when we consciously permit the influences, which most favorably mold our character, to weaken their hold upon our lives. Certainly in our time religion is the essential agent by which character is molded. Any of us would be foolishly short-sighted were he willing to weaken the hold of religion upon his life for the sake of a scientific theory, the truth or falsity of which could have but little practical bearing upon his conduct. We must hold to religion at all hazards. We may, when circumstances so suggest, change our denominational allegiance. We may and often do interpret our faith quite at variance with the ecclesiastical body with which we are connected. We may constantly modify and develop our beliefs. But it is a pitiful life which has lost the staying and strengthening influence of religion. I believe this conviction is deep-rooted in the minds of our people and that it deserves the place it holds.

To a mind thus essentially religious the announcements of science often come as a shock. They seem to run counter to our deepest convictions. It seems impossible to us that both can be true. Sometimes the more we debate the questions the more contradictory they seem to become. Every good mind needs unity in itself. No clear thinker can be quite content when two distinct departments of thought are at sharp variance in his mind. He may pursue one of two courses. He may hold to one view with conviction and earnestness and look upon the other as essentially false. To many religious people all science that runs counter to their convictions is necessarily false. They label it pseudo-science and pass it by. If the word pseudo-science is unknown to them, they stigmatize it as rationalistic, or still worse as materialistic and let it go at that.

The other course is to have faith both in religion and in science.

Such a fair-minded man must ask himself, what is the truth in the matter? If the scientific fact is true it is to be believed. It may run counter to what we have believed before. It may seem at first entirely incredible. But when once he becomes convinced of its truth the clear thinker must not only accept it, but must accept all legitimate deductions from it. If it seems true to us we must believe it. Absolute demonstrable truth, except in the simplest of matters is almost unattainable. The best we can ordinarily get is a close approach to certainty, and with this we must be content. In many matters, indeed in most matters, we must trust the judgment of others who are better trained in a particular line of thought.

As to the truth of geology we are certainly wise to accept for the present the facts and principles commonly accepted by competent geologists. In biology, we should respect the concurrent opinion of important biologists. We must not assume that a few biologists who think as we do are right against the biological world, or that a few geologists who think as we do are right against the geological world. For theology, we had better go to the educated theologian. But when it comes to reconciling two of these and to catching the inherent correspondence between them, it is often likely that each of these groups of men is unable to see clearly the view-point of the other. Here lies our freedom. Here we must either think for ourselves or think with those wiser than ourselves whose opinions seem to us to ring true and to focus for us our wavering and uncertain thought.

Among students of animals and plants there is no longer any question as to the truth of evolution. That the animals of the present are the altered animals of the past, that the plants of to-day are the modified plants of yesterday, that civilized man of to-day is the savage of yesterday and the tree-dweller of the day before, is no longer debatable to the great mass of biologists. To older men hampered by the convictions of an earlier age this dictum of modern science may still be a little uncertain.

The working biologists of the world have no doubt. They differ radically as to what brought about this change, they dispute vigorously as to the rate of change, but as to the fact of the change there is no difference of opinion. Under these conditions the thinking man is out of joint with the times when he sets himself against the idea of evolution. He may be so immersed in other lines as to be indifferent to the problem; but when he is hostile to it, he marks himself as clearly against his day. Many have been against their day and have been right. Very great men have often been against the opinions of their times and have come to be leaders of the world's later thought. But ordinary men in ordinary times who think differently on a special subject from the specialists of the times are not very likely to be right. It is safe for most of us to accept as true an opinion on which specialists on that subject agree. It seems clear to me then that the thinking man to-day has in the matter of evolution a double duty. He must become reasonably acquainted with the theory that so largely affects all present knowledge, and he must wrestle with the theory until it no longer hinders the hold of religion upon his life. He may be perfectly sure that he does not clearly understand both, but he must get them into reasonable concordance before he can be quite at peace.

Truth is true no matter how it is acquired. There can be no doubt as to the essential truth of religion: its fruits proclaim its worth. There can be no doubt as to the essential truth of evolution; the clarity it has brought into the sciences is the evidence of the value of the conception. That it will persist in its present form, that it will be unchanged by later additions to our knowledge is of course unthinkable. It may be incomplete, it may be undeveloped, but so far as it goes it contains the truth. Under these conditions, how can we bring peace into our own mind? These two important provinces seem so often to be at variance. The difficulty may lie in one of two places. In the first place, each truth may be stated in terms so peculiar to its own subject as to convey no meaning to the student of the other branch. There is a second, and more harassing possibility. The same words may be used by students in each branch but each side may put a different significance into the terms. Then each believes he understands the other, when he really does not.

Our theology is man's interpretation of God's revelations of Himself as recorded in the Bible. Our science is man's interpretation of God's revelation of Himself in nature. Each is God's revelation, and so far as we have understood it, that revelation is of the utmost importance in our lives. Each has all the inherent short-comings of man's interpretation. Each has all the difficulties necessarily found in any stage of a developing understanding. We may be sure if we could thoroughly understand God's revelation of Himself as recorded in the Bible and his revelation of Himself as recorded in the rocks and the tissues of animals as well as in the body and mind of man to-day, there would be no difficulty. When we understand both completely, as perhaps we never shall, there will be no contradictions of any kind between them. Even now if we are firmly convinced that truth must be in both, there will be little difficulty in reaching a workable unity which will satisfy the present needs of the human mind and will not be so crystallized as to prevent a future growth. If, however, we hope to find a unity between a belief in evolution and a belief in the inspiration and value of the Bible, we must accept both of these in the terms of to-day. To reconcile a twentieth century statement of science with an eighteenth century statement of theology would be as absurd as it would be to reconcile a statement of twentieth century theology with eighteenth century science. Each century must restate its truths in terms of its own time. The truths may be at bottom the same through many centuries but to be clearly intelligible in any century they must be couched in the terminology of the age.

It seems to me if we are to understand, in conformity with the thought of the age, any particular book in the Bible, there are three steps through which we must pass. We must first ask ourselves the kind of people to whom the book was originally written. We must know their habits of life and of thought. Until this is clear in our minds the book can have little significance. Having built up as nearly as may be the life and thought of the time, we must next decide what is the inherent truth taught to the people of that time by the book under consideration. Much that is written must be simply the setting in which alone that truth could reach them. This extraneous detail gives vigor and color to the message but is not the message itself. The last step and the hardest one to take, the one that to some minds seems almost irreverent, is to decide the form that message must take to-day to convey to our minds the same truth which the original message conveyed to the people of its time. In so far as we succeed in taking these three steps, we shall get the true message which this book holds for us to-day.

When Paul in his first burning letter told the Corinthian congregation that their women should be silent in their churches, he is not, it seems to me, giving a message which in those terms applies to the world to-day. If a woman has anything that is worth saying she has a perfect right to say it in church. In any denomination in which religious observance is not ecclesiastically formal she will be allowed that privilege. By an interesting peculiarity of mind on our part she may be permitted to do so upon Wednesday evenings, when our early prejudice still prevents her speaking on Sunday. What is the truth of the teaching of Paul in this matter? The Christians of Corinthian times had already begun to suffer from persecution. They were already despised and distrusted. Men had come to speak ill of them. Paul's injunction concerning the silence of women in churches was simply an injunction against their doing those things which in the thought and habit of those times were associated generally with looseness of character. Fine Corinthian women did not speak in public. A woman who would consent to speak before a group of men of Corinth of that day would by that fact have proclaimed herself a woman of loose morals. Paul's injunction is that, in this desperate struggle Christian women should do nothing which could possibly bring them into disrepute. The lives of Christians must be above suspicion. This message is certainly as true to-day as it was in the time of Paul and Corinth. Whether or not a woman speaks in church to-day has no bearing whatever upon the question. The question is how she speaks and what she says. If her life gives force to her message and her message contains God's truth she is entirely free to speak.

In similar fashion we have changed most beautifully the message which we have come to love, as the Mizpah message: "The Lord watch between thee and me while we are absent one from the other." We have absolutely transformed and glorified the message. It was once the calling down of the wrath of Jehovah upon one or other of two herdsmen if either of them should fail to comply with the agreement to remain within his own boundary. These men whose herdsmen were constantly stealing each other's cattle agreed to separate because they could not live in unity. They set up a heap of unhewn stone, and called upon God to guard and to see that neither of them passed beyond the boundary of the other. What was once a threat between warring herdsmen has become a binding link between Christian brothers. No longer do we call upon the Lord to guard in our absence lest our enemy encroach upon our domain. Now we call upon him to bind our hearts together so that neither time nor circumstance can bring division between us. The menace of a herdsman's wrath has become one of the tenderest messages of Christian love.

In the light of the principles stated above, what is the essential truth that lies back of the earliest chapters of Genesis? First, that there is one God. Slowly it had been borne in upon the Hebrew mind as upon no other tribe in the world that the Lord God is one God. Nearly all the world besides believed in many gods. Each nation had a God peculiarly its own, each city had a minor god caring for it particularly. There were gods of the woods, gods of the oceans, gods of the streams. Gods and goddesses were everywhere. To this people wandering through the terrible monotony of the sandy desert, the "Garden of Allah," there came the inspired comprehension of the eternal oneness of Almighty God. First, he was to most of them the God of the Hebrew, stronger than the gods of the nations. After a while under the teaching of prophet after prophet there finally came to the entire nation the exalted conception that God is one and there is no other God. This is one of the imperishable revelations of all time. Beside this, all suggestions of fifth or sixth day, of hours or of ages are absolutely insignificant. These are but the clothing of the idea which makes it acceptable to its time. This clothing must change with every age if it would reach thoroughly the minds of the age. Underneath and forever lies the glorious truth that the Lord God is one God.

The second truth which seems to me to underlie this magnificent parable of creation is the truth that this great God has created the universe and that he cares for his people. Gods before had been objects of terror. Gods before had lived lives such as the people themselves would not have respected among their companions. Gods before were to be shunned. If one could but escape the attention of the gods it was his greatest good fortune. Now we have the conception of an all-knowing, ever-present God to whom his people are dear. The terms in which it was stated in those days matter but little. To modern psychologists even the idea that people are dear to God seems speaking too humanly. Yet the truth involved must come in terms that the people of to-day understand. We can best comprehend God if we think of Him as loving and chastening, even though down in our hearts we know that these terms are not high enough, are too human to apply to an Eternal God. But we know no better and they tell us the truth even though the terms may in time pass completely away.

Last of all and perhaps most characteristic of the Hebrew people is the great lesson that this Eternal God, who created the universe and cares for his people, demands righteousness of his people. To the nations round about religion was not a matter of righteousness. For them religion had nothing to do with morality. Thieves might have gods favorable to them quite as well as righteous men. The worship of Diana of the Ephesians or of Astarte in the groves of the Asia Minor coast could be so unspeakably licentious and vile as not to admit of description to-day. Yet this was all religion. To the Hebrew came the inspired, exalted conception of a God who demanded righteousness of his people. Beside this wonderful revelation to the human mind details of serpents, and of apples, of names of men and of women, of gardens and of swords are absolutely but the transitory clothing. This brought them to the minds of the times. The value of the form is evidenced by the fact that it brought the conception. But we must not lose the glory of the conception in an over regard for the clothing in which the idea came.

Does this mean that Genesis has served its purpose and is to-day to be conceived of as a beautiful relic of the past, to be reverently enshrined but not seriously accepted? Far from it. The glory of the Genesis story lies in its wonderful power to grow. It strengthened the minds of a persecuted tribe wandering in the desert who finally settled in a small and barren country. It brought the truth to them so clearly that they have persuaded much of the world of that truth and bid fair to persuade the rest. The story has grown with the mind of man. As it served the Hebrew in his time it has grown to serve others to this day. Each generation has read the story in the light of its own times and each generation will continue to read the story in the light of its advancing knowledge. The only part of the story that can be affected is the clothing, the inherent truth remains forever. Furthermore, the story which persuaded the childhood of race is the story which will persuade the childhood of to-day. In no other form could the great truth of the Bible be brought to our children as well as in the form of these early chapters. In early life our children will accept these stories as literally as the ancient Hebrew accepted them. As they grow in knowledge, unconsciously and without jar, if we do not jar them, our children will read into the story what God has taught them in the world outside. The shock which came to their elders need never come to them. It is our fault if our children are disturbed by the conflict between religion and science which disturbed us. There is no difference between God's revelation of Himself, as we have it in the Bible, and God's revelation of Himself in nature. The better we know the Bible and the better we know nature the clearer this will be to us.

Perhaps the most severe shock that has come to the mind of religious man from the teachings of science has been the at first almost unsupportable idea that man is the descendant of creatures of which the ape is to-day the nearest representative. He had learned from Genesis the altogether adorable conception that he was made in the image of his Maker. It lifted him; it strengthened him; it gave him more power to struggle. He might know that he had marred that likeness by wrong-doing, he might understand that the fullness of the glory of God's image could not shine through his own face. Yet he believed that he was, in spite of all his imperfections, made in the image of his Maker. Now comes this horrible linkage with a miserable brute to either shock and confound him or to degrade him. We can easily imagine, some of us have bitterly experienced, the shock of this changed conception. But it was only because we mistook the clothing for the truth in both cases. We read science in its own terms; we read Genesis in its own terms. They did not use the same language and they jarred us to the very soul. Slowly, however, we are coming out of the darkness of that battle; slowly the glorious light of the beautiful truth is breaking into our minds and our hearts.

Michael Angelo painted a wonderful picture of "The Judgment." Here, seated upon a throne, which after all is only a magnificent chair, sits a venerable figure of what is really but a nobly-proportioned man, to whom the nations come for their final reward. He separates the righteous from those who must forever be sundered from their God. Seen through the distant past it still remains a majestic picture; but no painter would think of repeating its conception to-day.

Quite in the modern spirit is the beautiful lunette which John Sargent placed in the Boston Library, above his well known frieze of "The Prophets." It represents "Jehovah confounding the gods of the nations." The naked figure of suppliant Israel stands before an altar of unhewn stones, on which burns the sacrifice. The smoke ascends to Heaven. On one side stands the mighty figure of Assyria with uplifted mace ready to strike its awful blow upon the shoulders of helpless Israel. On the other side the lithe, subtle form of Egypt, clasping the knout, watches its chance to bring its treacherous thong upon the helpless shoulders of suffering Israel. But Jehovah may not appear, man may not look on God and live. Jehovah is seen as a glory behind the cloud of smoke shrouded by winged cherubim. From one side of the cloud comes a mighty hand meeting with power the force of Assyria. From the other side, a lithe and sinewy hand thwarts the subtlety of Egypt. But Jehovah is behind the cloud.

Again we understand that we are made in the image of our Maker. Again we understand the power of the uplift of this idea. From the conflict it has emerged in new and glorified form. Hath a God eyes that he may see? Hath a God ears that he may hear? Hath a God hands that he may work? These we know to be but human forms of speaking. Eyes, ears, and hands we may owe to the brute from whom we have sprung; in our eyes and ears and hands we show the relationship we bear to them. These are not the image of God. God is a deeper, a finer, a nobler something than hands, than ears and eyes. The image of God lies within ourselves: the image of God is that which makes us what we are. In every noble purpose, in every earnest endeavor to uplift ourselves or our fellowman, in every thought that turns us from the evil of a repented past, in every desire with which our hearts yearn to strengthen, support and sustain our friends and even our enemies, shines forth the image of Almighty God. This it is that links us with the Eternal: this it is that makes it worth while that we should be Eternal. Besides this what are hands and ears and eyes? We are made, all in us that is noblest and highest, in the image of our Maker.

A word in closing. The time is ripe for a broader conception of theology and of science on the part of those who are not trained to be specialists in either. We are becoming more and more inherently religious. We are becoming more and more enamored of the truth in all its forms. The times are ripe for us to cease the struggle and to strive for peace. So long as men insist that the important things in faith are the things on which men differ there will be eternal strife. So soon as men endeavor to find the common ground between them and each tries to state his belief in forms acceptable to himself but involving no hostility to his neighbor, we shall be working for peace.

Some of our finest men of to-day are being trained in modern science and in modern theology. There is no scorn in their minds for early science or for early theology. Each served its age, and each taught its truth. But its truth must be restated in terms of to-day. The old creeds will always be loved. The old creeds will always hold our reverence and allegiance. But each age must be at liberty to interpret these creeds in the terms in which that age best understands truth. Each age must be at liberty "to restate the doctrines of the past in accordance with the newness of the age and with the ancient verity of truth." How feeble my own attempt is in this matter, I quite understand; I am still a child of the struggle. It has all come in my lifetime and I have seen and felt not a little of the bitterness of it. I believe the time is ripe for a definite peace. I believe our children, if we do not hamper them, will never know the struggle we have had. In every great institution throughout this broad land men of earnest mind and noble soul are teaching the truth as God gives it to them to know the truth. Let us not hesitate to entrust our children to their hands. To us they may seem to be teachers of discord but they are not speaking in terms that we understand. They are using the language of a new age. Underneath their teaching lies the everlasting truth. Out of their teaching will come everlasting life. Let us trust God in the world. Let us believe that in this age he is teaching men's lips and dwelling in men's hearts. Only so can we give to our children the best their times can give them. If we insist in holding these men back to our conception we but deny them the privilege of moving with God's great procession. We make them laggards when they should be in the front ranks, their faces lighted by a nearer and clearer vision of Almighty truth.

###

www.ingramcontent.com/pod-product-compliance
Lightning Source LLC
Chambersburg PA
CBHW070858180526
45168CB00005B/1869